CLASSIC and Vintage COMMERCIALS

on

BEDFORD

Published by
KELSEY PUBLISHING LTD

Printed in Singapore by Stamford Press PTE Ltd.
on behalf of
Kelsey Publishing Ltd,
Cudham Tithe Barn,
Berry's Hill,
Cudham,
Kent TN16 3AG
Tel: 01959 541444
Fax: 01959 541400
E-mail: kelseybooks@kelsey.co.uk
Website: www.kelsey.co.uk

ISBN 1 873098 61 8

INTRODUCTION

'You See Them Everywhere.' Bedford's much-quoted classic advertising slogan of the 1930s still applies in some parts of the world, 17 years after the last example was built.

Britain's armed forces still rely on the marque in the new century. So did almost everyone in the country when 1950s Bedford RLH 'Green Goddesses' were pressed into action during a recent firefighters' dispute.

Bedford's beginnings date back to the mid-1920s when American company General Motors took over Vauxhall and started to assemble Chevrolet LQ and LT lorries in Britain, firstly at Hendon and then Luton. These sold well and, in 1931, the range was completely revised and sold under the name Bedford. These would now be very much a British product, though the first models used a Chevrolet cab.

Success was major and immediate, Bedford becoming a major player in the 2-3 ton market with its WHG, WLG and WS models and the range, joined by a BYC 12cwt van in 1934, proved rugged and reliable and deserved every bit of its success. Bedford also made major inroads in the bus and coach market, a situation that would remain for decades.

In 1939, Bedford announced its new range, but war got in the way and the 1.5 ton K, 2-3 ton M-Type and 4-5 ton O didn't reach full production until 1946. Meanwhile, the company had produced some 250,000 military vehicles, including tanks. In 1950, an S-Type seven tonner, known as the Big Bedford, was introduced along with an eight ton tractor. Wartime experience of building 4x4s was employed to produce the R-type, which also sold well to civilian customers. Not surprisingly, by 1954 Bedford had outgrown its Luton factory and a new plant opened at nearby Dunstable

In 1960 came arguably Bedford's best-remembered range, the forward control TK. A bonneted TJ, the ultimate in rugged simplicity, won much business in developing countries and was another long-term production survivor. Vans continued to bring in major business and Bedford was active in the larger lorry field with the KM range and the TM of 1974 gave Bedford good ammunition in the fight against Continental heavyweights.

The 1980s began on an optimistic note with launch of the TK-derived tilt cab TL and soon Bedford was taken under the control of General Motors' worldwide Truck and Bus Group, but a mixture of recession and steep decline in demand for vehicles – plus GM's failed plan to buy Leyland – led to a decision to pull out of the UK truck market in 1986.

Bedford may have gone, but there are many examples in preservation, some in regular use. Not everyone would describe Bedfords as glamorous – they would never be the stuff of Hollywood movies such as Convoy or Duel – but because Bedfords were cheap to buy, well-built and reliable, the marque's impact on Britain's haulage industry has been massive. Hauliers who began with a couple of examples often expanded into major players, buying British heavyweight lorries of other marques.

Nick Larkin
November 2003
Classic & Vintage Commercials

CONTENTS

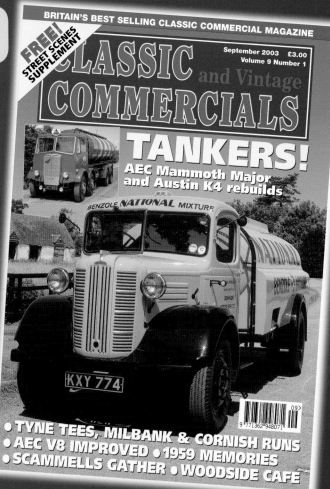

STILL IN DAILY USE

Our technical advisor Norman Aish paid a visit to Bob and Jill Foot down on the Somerset/Dorset border, for an in-depth look at their 1950 MST Bedford.

With the kindness of Bob and Jill Foot I spent a morning with them in early May. Bob and Jill run a timber business cutting and supplying logs. Their annual production is some 400 tons, which they deliver within a radius of some five miles of Yeovil, on the Somerset and Dorset border. Nothing very remarkable except that I have to tell you their sole working commercial lorry is a 1950 Bedford MST 3 ton lorry, registered MYB 78.

Jill's father started the business after the Second World War. In 1949 he commenced buying timber from the local Melbury Estate, and this source still provides the bulk of the timber today. Jill joined her father in the business in 1963 and it was in 1967 that the MST Bedford was purchased in a curious way. It came about that a radiator was needed for another MST Bedford that was being used by Jill and her father at the time. The MST

1950 MST Bedford of Bob and Jill Foot shows off its looks in early May 1995, note the towing shackle for pulling the lorry out the woods in soft ground.

Jill Foot who has tremendous skill which she has learnt over the years from her father unloads the Bedford with the 250-300 year old oak.

Fordson Power Major with fore-end loader, which is used for pulling out and loading up the timber which is taken away in the Bedford MST lorry.

stands for M model, S for short wheelbase, T for tipper by the way. The local scrapyard at Stoford had the radiator required but they would only sell a complete vehicle. After the usual negotiations which included the fact that a sledgehammer was included in the sale, the princely sum of £17.00 was agreed for

MYB 78, the 1950 Bedford. Apart from a new engine in 1985 and an engine rebuild in 1994 with parts supplied by Bygone Bedford Bits of Parkstone, Poole in Dorset, the lorry is out six days a week come winter or summer, though never on a Sunday.

The first job of my morning visit was to

go to Abbey Hill Showground at Yeovil to help clear and bag up the 30 sacks of sawdust from the rack sawbench demonstration, which had been steam-driven at this ever popular rally over the VE Day weekend celebrations of 6-8 May. Bob and Jill's geese seem to like the sawdust for bedding and certainly the aroma of fresh pine was very attractive on a sunny morning. We then had a quick run home in the Bedford to see the timber that had been sawn up at the rally weekend for the replacement tipping body that is to be constructed for the MST in the near future, as nothing gets wasted here.

Coffee was soon on the brew and home-made cake on the table which started the reminiscences going. This enabled me to see their other vehicle, which happens to be a KC 30cwt Bedford which has single rear wheels fitted, and a different axle arrangement to the MST. Other than an engine tune-up and some bodywork repairs to the rear wings soon to be replaced the vehicle is ready for the road. Bob is to join the HCVS, (Historic Commercial Vehicle Society), so he can be assisted in getting the lorry registered with the DVLA at Swansea. It has to be said that this KC Bedford was bought as a wreck, though fortunately the cab was salvaged from the other Bedford MST that the family owned in the '60s. This was the Bedford for which MYB 78 was obtained for that replacement radiator, so things have come full circle. (NYB 12, the KC, has been rebuilt from the chassis upwards and a superb restoration has been done.)

The Bedford MST has just arrived in the woods at Melbury Estate, one can clearly see the hefty branch that has split and fallen from the tree.

Down in the woods the 1950 Bedford takes its load of logs with ease.

A rear view of the Foots Bedford collection with the 1951 KC 30cwt on the left and the 1950 3 ton workhorse on the right.

NYB 12 was originally a flatbed lorry in the fleet of BC Taylor of Hardington Mandeville near Yeovil. Later the vehicle was converted to a hand wound tipper but the body got sadly twisted in the early 1960s, after an awkward piece of tipping. The lorry laid abandoned for some 25 years and, following the full restoration, it has been signwritten by John Burton of East Coker. The signwriting is all in the original livery and I must say is a credit to

Timber awaiting seasoning - this will eventually be used for the new lorry body that is to be fitted to MYB 78.

Seen here on September 1 1990 in a very rough condition, Bedford KC NYB 12 when the vehicle was being collected, look at our next picture for a change in its looks.

Photo Bob Foot.

8

It won't be long now, all the KC needs is those new rear wings and a log book before it's back on the road.

the craftsman who carried out the job.

Enjoyable as all this nostalgic discussion was I had to remember that this was not a rally visit but a business day with probably the oldest working Bedford lorry in the country, *(still perhaps you know different, please let us know - Ed)*. So off we went to Melbury Park to see a fine 250-300 year old oak tree, which had lost one of its branches. Bob and Jill pointed out the branch had rotted inside out from the tree

Showing the signwriting on the cab door on the KC.

trunk so that is why the branch had fallen and needed cutting up. Chain saw out and in no time at all we were loading about a ton of timber for the short drive to the part of the estate where the saw bench was set up. The lorry can take a maximum load of just over 3 tons with the extension sides fitted.

When loaded the lorry returns 12 mpg which is very fair bearing in mind the hard work the lorry does in the woods, and the short distances it travels. Of course when I visited the scene the weather was nice and

The axe man himself, Norman Aish doing a bit of chopping! We mean wood.

sunny; nevertheless, in the winter it is quite common for the MST to get stuck in the mud. When this happens out comes the early 50s Fordson E1A Major or the 1964 Fordson Power Major - both are redoubtable work horses in the Foot's fleet, both are fitted with winch's and one even has a saw bench on the front of it. The other Fordson has a fore-end loader fitted which is used to move the heavier timber around and out of the woods. The Foots mostly deal in beech, ash, oak and sycamore.

Well, lunch time was soon coming around so we gathered up all the sawdust into bags, chopped and loaded the timber, and at this point I had to say my farewells to Jill and Bob, not forgetting the MST Bedford. Fortunately the Foots have no plan's to retire their old friend which must be good news for everyone who loves old vehicles really working hard for their living-long may it reign.

From the Norman Aish, Bygone Bedford Bits files.

From time to time I will be featuring Bedfords and other makes from abroad. Rob Romain set out from the UK with his family in an ex-Crosville 1951 Bristol MW bus fitted with a 6LW Gardner engine. He drove across Europe, Asia and Australia ending up in Auckland, in the North Island of New Zealand. As can be seen the Bristol has been fully converted to his family home with all mod-cons. As a lightweight support vehicle Rob bought an ML Bedford which had started life working for an asphalt company. The wheelbase was extended by 18 inches when it was converted to its present condition in 1981. A most interesting style of "gypsy" bodywork was also constructed as can be seen in our picture, which was taken on a rainy day in New Zealand. Rob and his family have informed me that they have no plans to return to England and are enjoying every minute of their time in New Zealand.

Looking rather different than when first made in the early 50s, the MW Bristol coach and the Bedford ML Gypsy style caravan are helping to make a better life for Rob Romain and his family in New Zealand.

NORMAN AISH

Celebrates 25 Years of
Serving Bedford Commercial Owners

Adrian Fisher of Classic and Vintage Commercials took a trip to Poole, Dorset, in early May where he visited the remarkable and highly-respected Norman Aish who is celebrating 25 years of serving vintage vehicle Bedford owners world-wide.

Many people are fed up with these new-fangled items of equipment found under the bonnet of their car or truck, in fact they are an electrician's nightmare, or delight, depending which way you look at it. Generally there is a hankering after the more straightforward style of engineering of yesteryear. This nostalgia leads us either to admire old vehicles or hopefully actually own one. While some people are involved with clubs or partnerships many enthusiasts just work single-handedly on their vehicles.

Now this may be fine when everything is going well but what happens if you own a heavyweight and some large component needs attention. This may require proper lifting tackle and facilities so much as I

Norman Aish the man himself, who is just about to celebrate 25 years of business with Bygone Bedford Bits, seen here at the recent Dennis centenary celebrations at Wroughton airfield in Wiltshire.

admire AEC, Bristol and Leyland, there is something to be said in favour of lighter vehicles like Bedfords which can be worked on single-handedly that little bit easier.

So I recently took the opportunity to visit Norman Aish of Bygone Bedford Bits, the main suppliers of vintage Bedford spares for the 27 and 28hp six-cylinder petrol-engined machines in the UK. Norman tells me the story really started in 1968 when he purchased a 1952 Bedford

Reconditioned water pump rebuilt with a rotor rubber and tufnol seals and a new spring all made to special order. Photo Kim Henson.

Reconditioned clutch pressure plate assembly for the 28hp models built up and ground to profile. Plate re-machined, new springs fitted, also fork levers set up in a special press to ensure equal height. Photo Kim Henson.

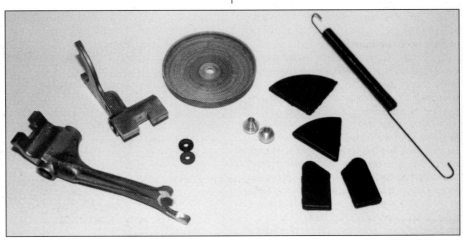

On the left is a gearbox gear selector fork built up with special alloy and machined. In the centre is a servo leather washer and two atmospheric rubber valves with conical round door hinge balls. On the right of the picture you will see another hot line in the form of bonnet corner rubbers and a clutch pedal return spring. Photo Kim Henson.

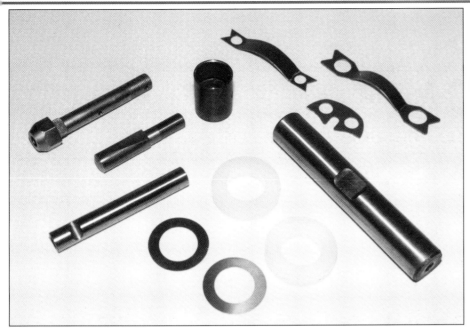

On the left of the picture we have a shackle pin and cotter. In the centre is a king pin and bush, main bearing lock washer again made to original specification. Photo Kim Henson.

MLC bus which can be best described as a smaller OB Bedford coach. The MLC is fitted with locally-made Lee Motors bodywork and the vehicle worked for Dorset County Council Education Committee as a school bus for some 16 years.

After a couple of years with the bus Norman thought that to keep the bus on the road it would be necessary to have a small stock of spares. As he went around the local dealers he was surprised to find so many parts available and it seemed a pity to see these parts go to the scrap man. It was not long before other enthusiasts asked why he needed to keep, say, six fan belts, and could he not sell one to them. So without any intention of going into business he found a ready demand for his stock. So this is how Bygone Bedford Bits came to be launched.

Good news seems to travel fast and soon he had owners from all over the country asking for assistance. Kind friends, especially Owen Angell who at that time worked for Vauxhall Motors in the spares department, gave Norman information about further sources of spares that were available and soon considerable quantities of parts were obtained.

In my visit I was most impressed with the 2,000 lines Norman carries at all times in his stores department. He not only sources his parts from the UK but he now purchases parts from countries such as Argentina, Australia, Belgium, France, Holland, Portugal and Singapore to name just a few. This has had to happen as many of the UK sources have now dried up, so Norman has had to broaden his horizons to keep up with the demand. A substantial number of items are now made to his special order by specialist firms which amount in total to some 20, so Norman is helping to keep British industry alive. These items that are specially-made include valve and clutch springs, servo leathers, brake rubbers and hoses, king pin sets and shackle pins to name but a few. The main criteria are obtaining parts of the right quality and at a sensible price. Tooling costs are always the biggest problem, Norman told us, together with the problem of finding engineering companies willing to make small batches.

Reconditioning parts is a most important aspect of the business. Mervyn Annetts, well known as the proprietor of Mervyn's Coaches and who uses his Bedford OB in his everyday business, does much of this work. The range includes overhauling to factory standard clutch pressure plate assemblies, oil and water pumps. Norman has other specialists working for him on electrical and ignition spares.

Most dealers and motor factors for modern vehicles cannot afford to keep too much stock because it either costs too much to keep on the shelf or it takes up too much room. So often dealers have to obtain items from the factory so customers face a delay of some kind. The supplier of vintage spares often holds many years of stock as it is often the case that he must buy stock just when it is available; it is a matter of being in the right place and the right time.

Norman's aim is to supply 95% of all mechanical and chassis components of the 28hp models from 1938 to 1953 and say 45% of all parts for the earlier W models fitted with the 27hp engine, with most of these spares in stock all the time. In addition to the range mentioned above Norman does keep a fair amount of stock for the A, S, and R model Bedfords as well, and he is always willing to source spares for these and other Bedfords from his many contacts world-wide.

Norman supplies spares to all parts of the world, although clearly the UK is his most important market. He operates an efficient dispatch service offering a 24hour mailing service and indeed same day dispatch if so desired, using Parcel Force, Red Star, Securicor, Royal Mail to name but a few.

As I drove away after my four-hour stay I could not have been more impressed with the dedication and efficiency of Norman Aish and Bygone Bedford Bits. There is no other comparable service available to a 'one make' in the vintage commercial world. If you are a Bedford owner, at least you have the knowledge that when you rebuild your vehicle you won't have to spent countless hours searching around the endless autojumbles in search of that elusive part; at least you stand a 95% chance of finding what you need at Bygone Bedford Bits.

Bygone Bedford Bits can be contacted at 11 Otter Road, Poole, Dorset BH15 3NH (Tel: 01202 745117).

Special Bedford QL oil pump with extended pick-up and filter arrangement with the more usual 28hp pump beside. All rebuilt with new impellers, relief valve piston and spring. Photo Kim Henson.

Carnival Time
IS HERE!

Norman Aish takes a detailed look at the unique Frome carnival float.

Through the kindness of Bill Ellis, Douglas Chandler and Bert Barton, I have seen what must be one of the most unusual Bedfords ever built.

were then constructed to widen the vehicle. The whole lorry was decked to create a length of 26ft 10in by 8ft 3in wide. The tyre equipment was changed from the military pattern to civilian to reduce the height to a remarkable 3ft 7in! The sides of

The 'mystery' vehicle under the out-of-season covers.   Photo N Aish.

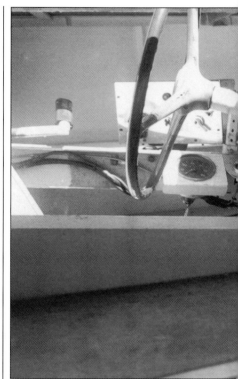

A sideways view of the driving controls; the gear handle can be seen on the left and the driver sits between the chassis extension members. A standard Bedford steering box is used but a universal joint and an adjustable frame are fitted. This enables the steering wheel to be raised when the higher driving position is in use on the open road.   Photo N Aish.

It started life in wartime as a water tanker, reputedly in the Western Desert, so it is assumed to be an OY 3 tonner. It was later used as a water tanker by Roads Reconstruction when they had a contract to build the Gloucester section of the M5 motorway. Bill Ellis purchased it for £30 apparently because it had a broken crankshaft but, in fact, it had a loose flywheel. Bill had plenty of experience with Bedfords because he was a fitter with Roads Reconstruction of Frome, Somerset, who ran a fleet of 90 Bedfords at that time.

The idea was to create a vehicle of the maximum possible size with a level deck of the lowest possible height throughout its length. The biggest problem was to devise a scheme to remove the cab and reposition the engine and, at the same time, still make it legal to drive on the road.

The first job was to remove the cab and engine and extend the chassis by 4ft 3in at the front and 2ft 7in at the rear. Outriggers

The covers are off and Douglas Chandler manoeuvres his way into the driving seat.
Photo N Aish.

the vehicle were fitted with valance panels to improve the appearance. So we now have a carnival float with looks like a large shoe box!

Compared with the next stage of the operation what has already taken place is fairly straightforward. The engine was to be repositioned between the axles, still of course in the vertical position. This was similar in style to the Bedford YRQ passenger chassis. The driving position seems to have been inspired by the Foden crane chassis by being dropped down in front of the vehicle.

TAM 897 of the Frome DAK Gang ready for the road - quite a machine don't you think. Photo N Aish.

The mechanical arrangements had to be modified in several ways as can be imagined. The large OY radiator manages to cope with the restricted air flow but, of course, it had been used to the heat of the Western Desert. The carburettor air cleaner has had to be specially-made to reduce the overall height. The carburettor could almost be gravity fed but the fuel pump has been retained. The accelerator is cable-linked using modified motorcycle parts. The clutch has also been changed; it is now hydraulically-operated so it is just a shade slower than the former direct mechanical linkage.

The crash gearbox has been replaced by a synchromesh box because it is easier than trying to match the engine revs with the gearbox so far behind the driver. It also solves another problem that the crash gearbox has a trigger release to enable reverse to be engaged and the syncromesh box avoids this problem. As it is, gearchanging is by a handle in the cab attached to a long rod which operates a double hinge plate. The inboard side of the plate has a track rod end connected to a shortened gear lever. My friend John Poole, who has driven the vehicle, states that clean gear changes can be made despite the length of the linkage. The propshaft is quite short as a result of the repositioning of the mechanicals.

The brakes are operated in the usual way with a Clayton Dewandre vacuum-assisted Lockheed hydraulic system. However, one important modification is to allow for the safety of people riding on thc float. Bedford brakes are capable of performances up to one G(the force of gravity). To give a much lighter brake performance when in the carnival procession there is a valve so the brakes can be operated on a mechanical hydraulic system only.

Not every vehicle offers the choice of two driving positions! Access to the driving seat is through a removable panel in the decking. In the carnival procession the driver sits below the chassis with the steering wheel in front and the column angled away to the offside through a

The 28hp Bedford petrol engine is repositioned between the axles. The modified air cleaner is situated on the right-hand-side. **Photo N Aish.**

universal joint. The gear handle is on the left-hand side together with the electrical control panel. The vehicle is glazed across the whole width so acceptable visibility is available for slow speed use. However,

In 1969 the Bells of St Mary's was the featured exhibition. We are looking at the front of the carnival float. Photo Bill Ellis.

A more recent display with the Bedford has been Santa Claus and his reindeer, with the front of the float on the left of the picture. Let's hear about your old carnival float. I am sure there must be other older or odder floats around; we would be delighted to know. Photo Bill Ellis.

whenever the float goes on longer runs to attend other carnivals, the seat is raised so the driver sits up higher with visibility over the top of the decking.

TAM 897 is a remarkable vehicle and at the moment it is used only at Christmas time when the Rotary and the Lions Club raise up to £5,000 for seasonal charities The 6kVa generator at the rear of the vehicle enables the display to be well-illuminated. While writing, I would mention that other south west towns have carnival processions in November with some floats up to 110ft long! Bridgwater, North Petherton, Glastonbury, Wells and Weston-Super-Mare are especially recommended with visitors from mainland Europe being attracted to the spectacle.

PICTURE ARCHIVE

The final incarnation of the Mercury in 1974, by then with 151bhp AEC 505cu. ins. diesel and Ergomatic cab.

KEEPING A BEDFORD ALIVE

Jim Treadgold tells Classic & Vintage Commercials all about his well loved 2 ton Bedford lorry which takes part in this year's HCVS London to Brighton run.

I bought my 1938 Bedford WH chassis number 4824 in 1966 from the late Brian Robbins who traded in commercial vehicles at the yard next door to us here at Wrotham, Kent.

Brian was a very good friend and neighbour. We both used to go to the WD sales at Ruddington, both for viewing and for the sales. We found this a very useful place to make contacts and have an informative time, as well as picking up the bargains.

I decided that I would purchase a historic vehicle to carry my stationary engines around to rallies. I mentioned this to Brian. He said 'have a look around the yard, I

should have something to suit your requirements'. I then started examining the vehicles up to two ton capacity and in the end it came down to either a Ford V8 short wheel base machine or a WH Bedford market gardener's truck. Both vehicles were built in 1938, however the Bedford won the day as it had a good Clement Butler & Cross body which was in reasonable order.

This Bedford was at one time part of a fleet of eight lorries owned by A Condon of

Richmond in Surrey, who was a greengrocer and fruit merchant. Condon's fleet comprised a number of Dennis lorries; one a 1927 5 ton machine, two three tonners and a pair of two tonners. These were complemented by three Dodge one-tonners and the Bedford, all mostly from the 30s. When Condon retired in 1958, Brian Robbins purchased the fleet and all survive in some shape or form apart from one Dodge which was made into a farm

Jim road tests the WH in preparation for the London to Brighton Road Run.

trailer.

A price was agreed on the Bedford and after much haggling I eventually paid the princely sum of £35. I towed the Bedford around to my property and started to get it in to serviceable condition. Interestingly enough the Bedford had been stored during World War Two.

The first job was to fit a new clutch centre plate, this was made easier by the fact that the gearbox had already been removed. I simply checked the gearbox over by removing the top, and to my relief all the gears were in good condition. However the floor in the cab was rotten so I replaced the required wood from stock items.

I then turned my attention to checking over the petrol engine. I replaced the plugs and the contact breakers, something one has to keep on the ball in any petrol Bedford. At the same time I checked the timing. All was going well so I connected up the 6-volt battery and primed the carburettor and petrol pump (both had been checked and cleaned). The moment had come to start the beast up, I turned the ignition on and with two or was it three swings on the starting handle she burst into life! It all sounded good.

I then drove the Bedford up and down the drive, when I found the brakes worked and appeared to be in reasonable order. However, the steering was found wanting as the track rod was bent. This was removed and straightened, greased up and refitted. When I came to remove the front wheels, the wheel nuts came off without any problem. However the wheels were stuck to the hubs and brake drums. The trouble

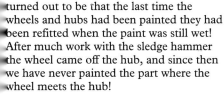

How it all started by using the Bedford to move our stationary engine about.

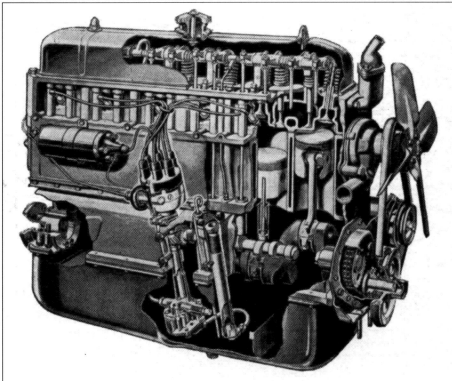

The Bedford engine which had a capacity of 3,519cc or just over 3½ litres.

turned out to be that the last time the wheels and hubs had been painted they had been refitted when the paint was still wet! After much work with the sledge hammer the wheel came off the hub, and since then we have never painted the part where the wheel meets the hub!

After checking out the mechanicals we turned to the bodywork. This needed some attention, particularly the paintwork which was touched up in places. The next task was to work out how to load the stationary engines into the back of the market gardener's bodywork. This was done with two thick wooden ramps fitted with angle irons at one end, that dropped over the tailboard. Also to make life easier a large plate with a very strong ring was welded to the front of the bodywork inside. This enabled us to use the block and tackle to load and unload the engines more easily, because at the time we had four engines to take around to the local rallies.

Our first rally with the WH was at Hadlow Down. On the way to East Sussex the 27.3 hp petrol engine of the Bedford stopped dead in the village of Five Ashes. Luckily my friend Ron Anstiss was following me in my Mini pickup. We checked the ignition for a spark and found the distributor rotor arm had cracked. So I stayed with the truck while Ron went to the nearest garage to find a new rotor arm. The reason I had to stay with the truck was that although I had taxed the vehicle; it did not come into force until the following Sunday! To top this we had come to rest outside the local police house and to compound things even more we were running on trade plates. So I smoked quite heavily while Ron was away finding a rotor arm! In those days one stood a very good chance of finding such things locally, and it was not long before Ron appeared with the required part. After fitting the rotor arm, away we went again and in great haste too!

On arrival at the Hadlow Down Rally we unloaded the engines and were asked to put the Bedford in the small line-up of commercial vehicles, which we did. I was very pleased with the old girl's appearance as she still carried the original Condon

The heavy duty, four forward speed gearbox was, at the time, a modern compact design, with short, rigid shafts to reduce whip and wear. Ball bearings throughout (except reverse shaft) minimised power loss and gave longer life. The top of the gearbox could easily be reached through a panel in the floor of the cab. The entire gearbox could be dismantled without dismantling the clutch.

The cylinder block is cast integral with the crankcase in alloy cast iron and the structure is stiffened by webs to give rigid support to the main bearings.

signwriting and lining. In the evening sunlight the cleaned and polished vehicle caused quite a lot of interest, remembering that the commercial scene was not what it is today. Our journey home on the Sunday evening was uneventful. This was to be the start of the Bedford taking our stationary engines to rallies in the south of England until 1970 when it was getting a little tired and needed attention.

By then I had been given a very interesting Blackstone lamp-start open crank 5hp engine. When it was restored I mounted it on a low trailer and towed it around to the rallies behind my Rover 2000. The Bedford took a back seat then and it was not until 1974 that work started on its second refurbishment.

By this time the floor in the bodywork needed replacing. I obtained the timber at a good price from a friend in the trade who had sympathy for old vehicles. I then treated the wood with Cuprinol and finished it off in gloss grey.

Now it was time to repaint the whole of the vehicle. This entailed stripping off all the

paintwork with one and a half gallons of Nitromors and lots of hard graft before we could see the original metal and wood. We had to wash the vehicle down well to make sure we had removed all traces of Nitromors. With this done we hand-painted the vehicle in its original colours of mid-Brunswick green (cab and body), red chassis and wheels and black mudguards.

As the vehicle had stood for over three years we found it would only fire up on three cylinders because some of the valves had stuck open in the cylinder head. We removed the rocker valve cover and gave the valves some manual assistance and gradually one cylinder after another came back to life. We decided we had better remove the cylinder head and decoke the valves, as well as giving the head a thorough overhaul. After re-assembly the engine sounded superb and was most definitely a six-cylinder machine again!

In 1975 the Bedford was accepted for the HCVS London to Brighton run. In those days there were not many Bedfords entered in the run, particularly like my October

A special point of interest in Bedford chassis frame construction is that all frames are riveted by then the latest 'cold-squeeze process'.

1938 interim style machine with the old style cab and the later type grille. Later on in 1939 would come the luxury of having the radiator cap outside the grille, not like my example where the filler neck is inside the grille and difficult to get at.

We had a trouble-free run up to Battersea Park, as we did on the run itself, as usual we were one of the first to arrive and had a fine time talking to other owners. However, the return trip was another story, as after the prize-giving we travelled via the Lewes-Tunbridge Wells route on our way home to Wrotham. The lorry suddenly came to a stop at Tunbridge Wells on Grosvenor Hill. We soon found that fuel was not getting to the carburettor so we changed the fuel pump in record time. It was a bit of luck we were carrying a spare fuel pump and in no time at all we were back on the road again.

The following year we decided we would sell programmes on the London to Brighton run. To this end David Hiatt and Peter Knight did the selling and I did the driving, so we were not an official entry that year. In those years we were attending many rallies in the south with the lorry, particularly the Kent County Show at Detling, where I had been chief steward of the commercial vehicle and stationary engine sections since 1967.

The spare wheel carrier fits neatly at the rear of the chassis.

MODEL WORLD

READER'S MODELS

Graham Vanstone gets stuck in with a Bedford-AWD local authority emergency support unit.

The Hart Models MK which gave Graham a head start with his MJ model as he gets down to drawing the bodywork.

The Bedford-AWD MJ that is still in 24 hour call out service and is the vehicle Graham based his model on.

I am sure most modellers would agree with me that when looking for a vehicle to model there is an extra bonus if the vehicle is both unusual and still active in service. The vehicle I chose is still on 24 hour emergency standby.

The one I decided to model is a Bedford MJ (displaying an AWD badge). The unit is an ex-military rigid 4 x 4 chassis/cab which was purchased in May 1993 by a local authority with a genuine 2,000 miles on the clock. Unlike the earlier Bedford MKs, this vehicle has power steering, also the 8,000 cc diesel engine received a much needed power boost from the addition of a turbocharger.

The current owners of the vehicle are Kennet District Council, who removed the more usual original canvas-covered metal lorry back and replaced it with an ex-military cabin. This was recovered from an

ex-Army Commer communications unit. Whilst this rear unit has been heavily modified, much of the cabin's earlier appearance still remains untouched. Internally, it has been completely restyled and equipped for its role as an emergency support unit.

The visual appearance of the vehicle in its new role and yellow livery is both eye-catching and interesting. But I was later to find that this vehicle represented a rather ambitious modelling project for a total novice.

When contemplating the best way of tackling the task in hand, I learnt from 'Transport of Delight', who are a well respected company in modelling circles, of a new model coming out. It was a high quality white metal kit of an Army Bedford MK which was to be released by an equally well known company, Hart Models of Hartley Whitney. Realising that the Bedford MK was not a million miles away from the later MJ, I decided to make use of this model and suitably modify the chassis and cab unit as donor parts for my model.

From a close inspection and comparison of the actual vehicle and model, it was clear that most of the modifications were

Below, the rear view of the Bedford-AWD showing the body detail - the body unit originally came from an ex-Army Commer.

PASTPERFE

By Peter Davies

O f all British lorries, Bedfords have dominated the preservation movement as indeed they did the transport scene for more than half a century.

Talk to any British haulier and the chances are that Bedford will enter into the conversation. So many transport operators started with a Bedford or, perhaps, learned to drive on a Bedford.

One such operator is Reg Jellis of R G Jellis & Son in Ivinghoe, Buckinghamshire. While to most hauliers Bedford's glory days are now just memories, Reg has kept the past alive. Such is the authentic appearance of his 1939 Bedford M type 2/3 tonner that it makes you wonder if time has stood still.

The 58-year-old truck captures the very essence of haulage in the early post war years. The distinctive frontal styling of Bedford's K M & O range is imprinted on the minds of most transport enthusiasts and there was a time when you could not go far without seeing one. Bedford's famous advertising slogan "You see them everywhere" was so true!

It is only in the past year or so that AAN 802 has been fitted with its period style livestock body, fulfilling 59-year-old Reg's long ambition to fill a gap in the preservation scene. "In this part of the country", he says, "cattle boxes were part of the everyday scene during the fifties and sixties when I started in haulage. I thought it was time that someone preserved a typical old-style Bedford cattle wagon. I spent about 14 years trying to find a Chamberlain body similar to the ones I used to have fitted, but they are almost impossible to come by. The body now fitted to AAN is actually by Lee Motors of Bournemouth and is the nearest thing I could find."

Reg has owned AAN 802 since 1968 and, believe it or not, he originally bought it with the intention of running it on haulage. "It was at the time when Plating and Testing was introduced", he explains, "so I thought better of it and decided to keep it original, just for preservation. At that time it was exempt from test, being

Inset: Reg – a haulage man through and through.

R. & I. JELLIS
Cattle Conveyors
IVINGHOE BUCKS · CHEDDINGTON 291

Bedford

AAN 802

Absolutely authentic – Reg Jellis' 1939 Bedford Two-ton cattle truck could easily be mistaken for a working truck.

registered before 1940."

As M types go AAN is a very early example, having been first registered on September 25, 1939. Reg pointed out from information he has that the engine number is in the earlier W type series. Another interesting point is that it has a W type radiator badge (blue background) as opposed to the normal red and silver M type badge. Wooden seat boxes and cab steps are also early features. This makes AAN one of the oldest M types in preservation even though, at first glance, it could be mistaken for one of the post war versions which are a lot more common.

AAN was found at a farm in Pitstone just a stone's throw from Reg's own yard at Cheddington. Farmers, J H Hawkins & Son, had owned it since 1954 and it had been engaged on general farm duties, including carting the pigs to market. The truck's origins are a little obscure but it appears to have been registered to an owner in East Ham, London, before being commandeered by the War Department for military use during World War II.

It was fitted with a high sided body, presumably for the transport of ammunition, and possibly had a canvas tilt which had gone missing during its latter years. Reg's mechanic at the time, Norman Worvell, who served with the Royal Army Service Corps, believes it was one of a large number which passed through the MoD's Carlisle paint shops while he was based there. In its army guise AAN wore an all-over coat of battleship grey drab.

Having bought the Bedford, Reg towed it home to Cheddington behind a farm tractor. Despite having stood for many years, it started up with very little trouble. "We drained the old petrol out and put some fresh in, put on a good battery and she fired up", recalls Reg. In the event full-scale restoration did not start until some five years later but by 1975 AAN had been completely refurbished and she was ready for the rally field.

The original engine was replaced as it

Latest addition to the Jellis fleet is this ERF EC-10 eight-wheel rigid, which looks superb.

Chas Jellis's superb Seddon Atkinson 411 reg., F119 UAV- an award winner at Truckfest.

It's an early 'M' type – the Griffin badge is as the earlier 'W' type.

was worn out, having been re-sleeved and subsequently rebored to plus 60 thou during what appeared, to have been a hard life. Reg still has the engine and plans to overhaul it eventually and put it back in. The restoration was very thorough as Reg is a real stickler for good maintenance and original detail. So often, well-meaning preservationists add their own touches to personalise their vehicles. Some decide to chrome their rocker covers and polish up bits of brass that would normally be painted. Instead Reg has kept AAN 100% authentic.

For the past twenty odd years the Bedford, in platform lorry guise , has attended scores of shows and rallies. Throughout that time Reg had been on the lookout for a cattle box body to complete the effect but it wasn't until a chance meeting with another Bedford enthusiast at the 1993 Woburn Rally that he struck lucky. Just the right period style body turned up mounted on a 1954 Bedford A type. While it was not a Chamberlain body

The authentic nameboard is actually from one of Reg's working cattle trucks from the sixties.

Above and below: Period-style bodywork is by Lee Motors of Bournemouth – the nearest Reg can find to the Chamberlain bodies he used to have fitted.

it had all the right characteristics, so Reg wasted no time in doing a deal.

The end result looks spot on and is a nostalgic reminder of the days when Reg first set out in haulage during the late Fifties. Even the name board and the RHA badge are 100% authentic. A nice finishing touch is the genuine late fifties B licence disc stating the conditions of operation.

The Jellis family history goes back to the 17th century in the Pitstone area and has early connections with transport long before the arrival of mechanically propelled vehicles. The Jellis's were farmers, butchers, millers and coal merchants. Reg's great uncle Fred owned the first petrol engined cattle truck in the area, a World War I Thornycroft. During the late Thirties when the Tunnel Cement Works was established at Pitstone, Fred hauled one of the very first loads of cement from the new works. Tunnel Cement was later to provide much of Reg's work from the late sixties onwards, and it was one of Reg's lorries that hauled the last load of cement from Pitstone on its closure in 1993.

Reg's father, Charles, was a stock farmer but ran his own butchery and slaughterhouse at Pitstone, founded in 1931. He bought his first Bedford back in 1933 – a BYC van. An ASYC followed in 1937. In 1949 he invested in a secondhand Austin K2 livestock lorry, followed by a second K2 in 1954. These were used to transport his cattle both to the market and to his own slaughterhouse at Pitstone. It was in 1953 that Reg joined his father to begin his long career in transport.

"Dad had taken delivery of one of the last Bedford PC vans in 1952", he says. "It was at the time when the CA was being introduced but he was a bit wary of buying a brand new design and preferred to stick with the traditional PC. In actual fact Dad had ordered the new PC back in 1939 but because of wartime restrictions he had to wait 13 years for it to arrive! I passed my driving test on that PC in 1954 and we still own it to this day." The old PC has been completely refurbished and put back into its original livery and even has its original scales and equipment in the back.

Reg's early experience on his father's lorries, hauling livestock and carrying farm produce to the London markets, stood him in good stead for his career in haulage. As well as rearing cattle his father owned a large prune orchard and the prunes used to be transported to Brentford and Covent Garden markets on a regular basis. "This part of the Chilterns between Totternhoe and Weston Turville was the only part of England where prunes were grown", explains Reg.

Having cut his teeth on his father's lorries, Reg decided he would try his own hand at transport in 1955. He bought an old 1936 Bedford WS 30cwt and spent weekends selling manure in the surrounding towns and villages, venturing as far as Hemel Hempstead. "I found I could make three times as much money doing that in a weekend as Dad could pay me for a whole week's work", he recalls.

An essential piece of kit is the paste pot full of petrol to prime the carburettor. Note the cattle stick making an ideal bonnent prop!

Four years later Reg began his own haulage business proper by investing his money in a newer O type Bedford 5-tonner with a B licence. R G Jellis & Son grew from there and currently operates a modern fleet of 38-tonne artics plus a new ERF 'EC10-350' eight wheeled flat, which is the latest addition to the fleet.

In the early years Reg operated a variety of Bedford, Commer, Dodge, Leyland and Thames four wheelers mainly on livestock, but he diversified into general haulage as the livestock trade dwindled. One of the main commodities hauled was, and still is, cement. Until 1993 it was produced at the nearby Pitstone Works of Tunnel Cement

and Jellis's long association with that company began in 1967. Following the works' closure in 1993, Reg's fleet continued to haul cement mainly from the Ketton Works in Leicestershire. Other goods hauled on a regular basis included timber, flour, animal feed and fertiliser.

The first move away from the cheaper mass produced trucks came in 1973 when an AEC Mercury was added to the fleet. Until 1976 the fleet was all rigids but in that year two Bedford KM artics were purchased. Soon two AEC Mandator units came along and from then on the Jellis fleet has been predominately artics. Regular visitors to Truckfest at Peterborough will

no doubt be familiar with the immaculate 1988 Seddon Atkinson 411 tractor unit from the Jellis fleet. That unit, F119UAV, is the pride and joy of Reg's 31 year old son Chas, who joined the firm in 1982 and has recently taken over the reins. In November '96 he and his truck took part in the Lord Mayor's Show, carrying a boat for the Wendover Trust.

Reg's wife Irene, herself a qualified HGV driver, keeps all the paperwork up to date and plays an important role in the running of the company. Reg himself packed in truck driving about ten years ago but still enjoys getting behind the wheel of his beloved M type. He has been taking things easier recently following a serious illness which landed him in hospital for two years to undergo a number of major operations. Thankfully he has made a remarkable recovery.

Having spent his whole career in transport he is a mine of information on the history of local hauliers and is a collector of memorabilia. The M type is testimony to his expertise in old vehicles – everything about it is totally authentic, even down to the paste pot on a bit of string. "That's most important", he grins. "It's vital, especially in the cold weather, to prime the carburettor before starting. You dip the pot into the petrol tank and pour a drop of petrol directly into the carb. With the six volt system it takes too much out of the battery while the lift pump is trying to pull the petrol through."

Another curious item of equipment is the long stick laid on the cab floor. "That's the old cattle stick," explains Reg. "Every cattle lorry driver needs one but it also has

The period enamel RHA badge adds a nice touch.

A genuine enamel Bedford Driver's badge adorns the front bumper.

Reg's lorry even has a surviving B licence from the pre-1970 era.

another purpose – it's just the right length to prop the bonnet open!"

Amongst the many transport bygones in Reg and Irene's house is one item that has puzzled them since Irene bought it as a present about ten years ago. It is an antique brass bottle opener in the form of a large door key and is engraved with the words 'Bagal Motor Service, New Delhi – Key to Vauxhall Bedford Service'. It is presumed to be a promotional item but if any readers can shed light on it, or date it, then Reg and Irene would be interested to know more.

Reg's considerable hoard of vintage Bedford spares has proved invaluable to a number of preservationists – such rare items as original WLG model carburettors and constant velocity joints for torque tube transmissions are among the Aladdin's cave of parts he has amassed over the years. Thanks to enthusiasts like Reg we can all enjoy seeing past transport brought alive in a real and meaningful way.

Reg's lorry is not mollycoddled for concours cups at rallies – he doesn't mind it smelling of cow dung – just like the real thing.

Back to Life!

Plant hire contractor, Tony Robards, with his Bedford TL that can tell a story or two.

The Editor visited Robards Plant Hire to look at a Bedford WS to be brought... Back To Life?

Well known plant hire specialist Tony Robards from Ringmer, East Sussex, has recently unearthed the remains of a 1934 WS Bedford 30cwt dropside lorry. Tony has had a fascination for old motors, particularly commercial vehicles, which goes back years, however, with commitments to his business he has never been able to indulge himself.

He was recently working in a disused farm in Ringmer when he came across a vintage lorry of some kind. Despite not knowing exactly what it was, Tony purchased it on impulse, and that's when the fun started!

Access to the site was difficult but, having cleared the brambles away, he was able to see what it was. At first he thought it was a Chevrolet LQ truck of the late 20s, however after a visit to the local library, he eventually decided the lorry was a Bedford. The single rear wheels indicated a 30cwt model and the engine number was found, which indicated the vehicle was built in 1934 and the engine could still be the original unit. Sadly, like so many Bedfords of the 30s, the chassis number is missing from the dash panel on the near-side.

Having collected the parts that had rusted away, it was time to bring in the mini-digger which was used to pull the Bedford on to some hard standing. After much effort, this mission was accomplished, and the vehicle was loaded onto Tony's beaver-tailed Bedford TL and brought back to his depot.

On close inspection, the vehicle is in quite a state. The front engine cross member has rotted away, as have many other parts. The engine distributor has obviously been removed in recent times, by the looks of things, and the Lucas starter motor housing on this 6 volt lorry has rusted right away! The engine is fitted with later type fuel pump and the Zenith up-draught carburettor, however, the block is of the correct type for the year.

When the vehicle was last used, one would have thought it was in reasonable order as, surprisingly, the steering box and joints are in good order. On the door frame is a Mansfields Ltd plate - they were Vauxhall-Bedford agents in Eastbourne, Lewes and Hove right up to the 1970s. It is presumed that the vehicle was supplied from the nearby Lewes branch, which was just up the road from Ringmer, but who

Where is the bodywork? It's there somewhere, but the elements have taken over! It's thought that the lorry has been in this position since the 1950s.

knows...

Obviously the engine is seized and, at present, Tony is attempting to free it off with Plus-Gas, lots of diesel and a fair bit of praying! Without doubt, the WS is in quite a state, but Tony is keen to restore this vehicle, particularly as he is a Bedford fan.

As you can see, Tony is in a dilemma over what to do with the vehicle, and is not sure which way to turn. However, he would be delighted to talk to other owners of WS Bedfords from this period. Please remember this model was introduced to the public in late 1931, production ran from 1932-35. The next WS series was introduced in late 1935, and was very much a new model and totally different to the earlier WS - it was to last in production until 1938, when the next styling change took place.

Tony asks how much it will cost to restore; well that's a bit like asking how long a piece of string is! Obviously the vehicle needs virtually every panel, besides many other things - do you have any panels that will start Tony off? Having talked to a Bedford expert from Cambridgeshire, he suggests that to restore the vehicle will cost around £16-20,000 in total.

He added that, when it's rebuilt, it will be worth only around £8-10,000, which is not a good prospect to a restorer, no matter what. Tony wants something good to come out of saving this vehicle. Please give Tony a ring on 01825 750472, he will be delighted to speak to you about the WS.

As can be seen, the panels on the Bedford are well past their sell-by date, do you have any better panels that will help Tony?

This Bedford just does not want to die... the steering wheel could be remade.

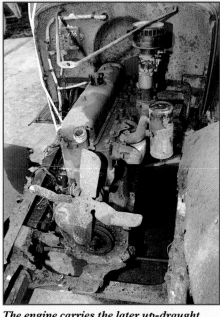

The engine carries the later up-draught Zenith carburettor; however, the block is of the correct type for the 26.33hp design.

What the end result could look like; this Bedford WS was seen at the old Crawley halfway point on the London to Brighton Run in the 1980s.

Main Pic: The vehicle, on a cold and windy Friday February 27 1998, the cab strapped up for the removal job.

The front cross-member has rotted out completely - does anyone out there have a pattern?

Tony Robards, plant contractor, checking the water pump bearings on the WS - Bygone Bedford Bits keep these in stock.

Significant dates in the history of the Bedford WS 30cwt

April 1932 - introduced with 26.3hp six-cylinder engine - no radiator stoneguard, torque tube drive.

June 1934 - September 1935 produced with Hotchkiss drive - open propeller shaft.

October 1935 - April 1936 - as above, but with forward-mounted engine, new panel work and vertical grille stoneguard.

May 1936 - October 1936 - as above, plus external spark plugs, different cylinder head and other engine developments.

November 1936 - September 1937 - as above, plus new steering connecting rod, now with an 'H' section forging instead of round rod.

October 1937 - June 1938 - as above, plus cooling system thermostat.

July 1938 - May 1939 - now with a 27.34hp engine, new radiator grille with horizontal louvres and an internal filler radiator cap.

June 1939 - September 1939 - the K model introduced, which superseded the WS - as above, but now with a pressed steel cab with divided 'V' sloping windscreen and external radiator cap.

Green Goddess RIDES AGAIN

A fraction of a photograph has resulted a unique reunion for an AFS fire crew, and the Green Goddess they used throughout their existence. Nick Larkin reports.

March 28 1968 was anything but a happy occasion for the smartly turned out Huntingdon crew of the Auxiliary Fire Service. This was the sorry day on which they, and other AFS brigades across the country, were forced to disband, by order of the Government.

The close-knit team hid their emotions to gather for one final photograph in front of the Bedford SH2 'Green Goddess' fire appliance, which had been with the crew since 1955.

As the team packed up for the last time, no one in their wildest dreams could have predicted that, 30 years later to the day, that photograph would be recreated, in the same place, with the same Green Goddess.

All the team are fit and well today, and it was only a holiday which prevented the group's training officer Maurice Johnson, who went on to become Chief Fire Officer of Oxfordshire, being there to complete the line-up.

It was only by sheer fluke that the non-human part of the line-up, a certain Bedford-registered NYR 727, was there too.

AFS historian, Richard Oliver,

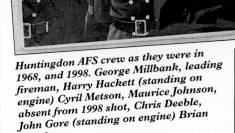

Huntingdon AFS crew as they were in 1968, and 1998. George Millbank, leading fireman, Harry Hackett (standing on engine) Cyril Metson, Maurice Johnson, absent from 1998 shot, Chris Deeble, John Gore (standing on engine) Brian Wrighton.

Green Goddess owner Gordon Newton.

discovered an archive photograph with just a glimpse of the vehicle's rear offside - and its registration number.

"I was sure the vehicle still existed, and made enquiries," Richard said.

The result was the Bedford's return to Huntingdon Fire Station for the first time in 30 years, and once again the

clank of its bell rang across the yard.

Brian Wrighton, who normally drove the appliance in service said: "We are all delighted to see it again. The Bedford was, shall we say, very mechanical to drive and did need to be handled with care, particularly when it was full of water. I was a lot younger in

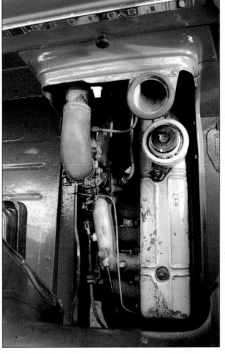

New Bedford 300 engine has been fitted.

No frills in Green Goddess cab.

Much effort has been made to re-equip the appliance with authentic equipment.

those days, though!"

Astonishingly, members of the team have kept in touch throughout the past 30 years, which must be a record of some sort!

The AFS crew enjoyed a good relationship with their full-time fire service colleagues. Training sessions took place twice a week, during which they would be invited to assist at incidents ranging from chimney fires to a lorry load of dog food overturning on the A1.

Set up in readiness for any nuclear attack, the AFS was disbanded when the Government felt it could no longer finance the service.

Many AFS volunteers went on to join the regular fire brigade.

So what became of the Green Goddess NYR 727?

After being in storage, the unit finished up working at Carlisle airport, in an indelicate shade of yellow.

It then passed through the hands of several private individuals, before being bought by the present owner, Gordon Newton, a couple of years ago.

Believed to date from 1953, and among the first batch of Green Goddesses built, the 4x2 spec appliance has been repainted inside and out by Gordon, who has also installed a 'brand-new-from-a-box' Bedford 300 engine. He has also spent many hours tracking down replacement equipment for the now fully-functional vehicle.

Gordon bought NYR up to Cambridge from Kent for the reunion, enduring its 5mpg fuel consumption.

'The AFS is the forgotten military force and, today, a piece of its history has been recreated.'

Asked to comment on his prize possession, Gordon sums up: "It's a fire engine. It's a Bedford. What else do we need to say?

BYGONE BEDFORDS

Martin Perry continues his bi-monthly series on older vehicles still on the roads of the UK - have you seen anything of interest for Classic & Vintage Commercials lately? If you have please send it in.

Whoever it was in Bedford's pre-war publicity department that coined the slogan 'you see them everywhere' cannot have ever envisaged how true that fact would become.

Looking back at my old 'spotters notebooks' from the 1960s, Bedfords figure prominently on each page. There are not many evenings when we can't sit back in the armchair and watch feature films from the '50s with a street scene and there in the background are Bedford O types and S types at work!

Watch any episode of *The Saint* or *The Avengers* and there on the back projection will be a TK or an A type. The same could be said of the 1970s series *The Sweeney* or *The Professionals*. Filmed on the streets of London, no episode goes by without plenty of TKs, CAs or little

Yes it is 33 years old and still hard at work, this 1965 TK operates for a farmer from the Herefordshire village of Thruxton and is seen here 'on the wash' at Hereford cattle market. It was on September 6 1960 that the TK range was announced to the world. That wooden varnished cattle float body is a rarity in itself, since in the '60s the lighter aluminium bodywork took over.

HAs going about their business in almost every shot!

There are, however, still a dwindling number of Bedfords, some now over 30 years old, still at work up and down the country, keeping the name alive. Once mundane in the extreme, they have become nostalgic reminders of a fondly remembered but fast disappearing marque.

This then is a photographic tribute to what still remains on the road carrying that famous name of Bedford, that were taken on the roads in my home county of Herefordshire during 1998.

Dating from 1974, this 466 engined TK lives in Kingston where it operates for Hedley Simcock on hay and straw haulage. Bedfords have been very popular with farmers and hauliers connected with agriculture over the years. When this example was built most bales in the UK were square. Since the early '90s that has all changed and the lorry now gamely copes with round bales (round bales first became popular in the USA when Allis Chalmers introduced their Roto-baler in 1947 - Ed.).

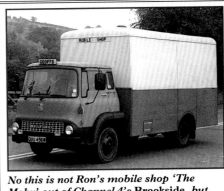

No this is not Ron's mobile shop 'The Moby' out of Channel 4's Brookside, but Geoff Morris's example which travels around the Bromyard area. The TK carries the Bedford four-cylinder 220 diesel engine, which the Editor rates as a very good engine, apart from the rear main oil seal design.

A very rare beast operating seen at Kingston is this well-used ex-MoD four-wheel drive Bedford MK. It carries an aluminium demountable tipper carrying wood waste for a local insulation manufacturer.

Another hardworking survivor, which remains in superb condition, is this 1970 TK which carries a six-cylinder 300 series petrol engine! The vehicle was purchased new by the Reese family at Credenhill, and is seen here unloading sacks of coal at Boddlestock, Hereford. The Bedford shows no signs of rust and is well maintained.

Bedford's final commercials were also their heaviest, with the TM range offering gross weights of up to 32 tons. This is typified by this 1986 Cummins engined 4x2 unit. Certainly the TM was a far better truck with a Cummins E290 engine compared with the less flexible GM Detroit V6 and V8 engines. The example seen here is owned by ABE from Ledbury, who ran a fleet of these vehicles for many years and liked them. Sadly we see the last in the line, D324 VMO. I photographed the TM last year on a stormy day at Bromyard loading car components for delivery to Swindon. An era is at an end.

CREATIV

U ntil four years ago, R C Jeffrey's smart fleet of Mercedes artics was a familiar sight up and down the country, but at the end of 1994 the decision was made to sell the family haulage business based at the picturesque village of Pebworth near Stratford-on-Avon. Since pulling out of haulage, Mick and Robert Jeffrey have turned their attention to lorry preservation and the results of their endeavours are now being enjoyed by rally goers and fellow enthusiasts.

The vintage fleet has already grown to eight vehicles plus another four trucks from the '70s and early '80s, which have been preserved but will also serve as transporters for the old timers. Mick Jeffrey, 55, and his elder brother Robert, 63, have spent their whole careers in haulage and took over the family firm from their father during the '70s. Robert was in charge of fleet maintenance while Mick looked after the traffic office. Both have always held an interest in old vehicles, but it's only in the past five years or so that they've been actively involved in preservation.

Bedfords hold special memories for the brothers since they were brought up with them. For about 25 years, up to 1970, Jeffrey's ran an all-Bedford fleet, including many O-types and S-types. It's therefore no surprise that their vintage collection consists of seven Bedfords spanning the 1936 to 1955 period, plus two TLs from the early '80s. Breaking the all-Bedford mould is a solitary Austin Loadstar of 1952 vintage plus two 'modern' vehicles - an AEC Mercury and a Mercedes Benz tractor unit - both dating from 1978.

The first vintage Bedford was purchased in 1994. This is the beautifully restored 1937 WLG 2-tonner BWP934 restored by Reg Jellis, the well known Bedford expert from Cheddington, near Ivinghoe in Buckinghamshire. The platform bodied truck was already in tip-top condition and

Vintage Bedfords dominate the Jeffrey's collection, but the Austin 'Loadstar' is also a fine classic of the '50s.

Austin 'Loadstars' are not as common on the rally fields as Bedfords. This one started life with the army but is now resplendent in Jeffrey's traditional style livery.

only needed repainting into Jeffrey's smart maroon and black livery.

Very soon a second, earlier WLG dating from 1936 was purchased from an owner at Melbourne, York. This is registered WF9851 and, like BWP, is fitted with a platform body. In their search for a variety of Bedford types, Bob and Mick were pleased to come across a restorable 1945 OWL 'square nosed' 5-tonner at Ledbury in Herefordshire, not too far from home. The wartime spec civilian 5-tonners are true classics, and FDD428, new to Williams of Banbury, is now safely garaged at Pebworth where it is undergoing a full restoration.

During the ensuing couple of years, more Bedfords were sought out and preserved. These included one of the ex-MoD OLBC tankers which turned up at L W Vass of Ampthill. This dates from 1951 and so has the 'extra duty' spec and synchromesh gearbox. This will be restored as a flat too and, when done, will

PERIOD

By Peter Davies

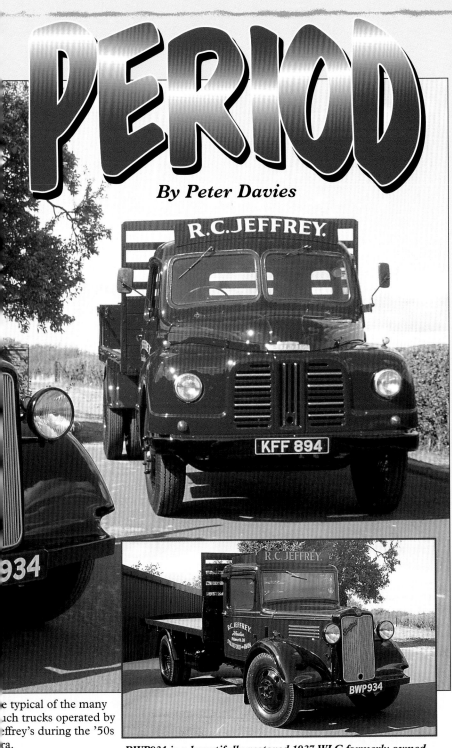

Platform body with tailboard and high headboard was typical of the market gardeners' lorries of a bygone era.

Austin is the odd one out in Jeffrey's virtually all-Bedford collection.

e typical of the many
ch trucks operated by
ffrey's during the '50s
ra.

Another ex-military
machine is a 1955

LC3, LYO776. This had been partly
estored and fitted with a dropside body.
The completion of the restoration work
as been farmed out to a restorer at Great
Yarmouth. The plan is to put it back into
he period Jeffrey's livery so that it looks
milar to TWP700. Fortunately a colour
hot of TWP was taken in 1960 after it
ad just been loaded to the gunnels with 9
ons of flake maize.

In 1996 came the amazing discovery of a
953 O-type, still at work with a local
ompany called Kinver Sawmills. This is
ne of those 'time warp' experiences that
ccur all too rarely. The truck instantly
riggers memories of the early post war
ra. In its unspoiled working state it's a
ngible living link with the past. Once

BWP934 is a beautifully restored 1937 WLG formerly owned by Reg Jellis.

restored, it will of course lose its priceless
'patina' that only comes with real work,
and I for one felt privileged to have seen it
and to have taken in the magic of such a
rare find.

The truck, a very late example of the
OLB with the A-type rear axle, was used
by Kinver Sawmills on local work right up
until February 1996 and was put up for
sale following the owner's death.
Amazingly it has only clocked 53,000
miles in 43 years and has survived in
perfect order.

Breaking the Bedford monopoly is a
1952 Austin 'Loadstar', KFF894, which
the brothers bought from Tom Howard
Banks at Southport. This is an ex Army
machine now resplendent in civvy guise

This 1953 O-type has survived in an all original state, having only been retired from work in 1996.

This splendid event was the second gathering of Bedford vehicles to be held at Somersham in association with the Cambridge Omnibus Society, set in the premises of Dews: The Coach Company. The town is situated north of Cambridge between Huntingdon and Chatteris. Following the success of last year's inaugural event, many more Bedford buses and coaches were present and the show also included a large number of classic and vintage commercials.

Of all the bus and coach types surviving into preservation there are probably more OBs around than any other type. There are around 180 listed as preserved and out of these some 40-50 are in roadworthy condition. The OB is still a very popular

Joe Ovel's 1936 Bedford BYC is powered by a 20hp six-cylinder engine.

Derek Styles seen here in the village of Somersham on the way to the rally site with his smart 1949 Bedford KD.

type at rallies, as evinced by the large number seen at this year's gathering. The Bedford OB coach has also proved to be very popular with some operators who provide heritage tours, private hire and weddings. In fact I had the pleasure of travelling on the free vintage bus service which operated throughout the day between the rally site and the nearby historical town of St Ives. A special welcome was given to the commercial vehicle owners who turned out to give everybody the opportunity to view their vehicles and compare them to more modern types seen on the roads today.

ERF of Sandbach have recently purchased the Bedford name from Marshall SPV of Cambridge, and one hopes that in the future we may see Bedfords rolling off the production lines again. A mission statement from ERF stated, "We are delighted to have taken over the Bedford parts range along with the Bedford network, which has now been integrated with the ERF network. Our aim is to improve the levels of support to the Bedford customers

Bedfords at

By Alec

The smart 1961 Bedford J1 dropside with owner J Light, who travelled all the way from Chingford in Essex.

to make sure we carry on offering the best after-market service."

Some 30-40 commercials turned out to celebrate 60 years of the Bedford OB - these included not only Bedford lorries but AECs, Volvo and ERFs, a truly splendid sight. I had the opportunity to photograph some of these fine Bedford commercials just outside the rally site in the village. This brought

back many memories for the owners, who used to drive these splendid lorries for a living. In the town of Somersham there is perhaps the only surviving garage with the original forecourt and petrol pumps; the "West End" garage dating back to the 1940s. The pumps are no longer in use, however, but the garage is still used for restoring vintage cars and motorcycles and

Somersham

tschuk

A splendid 1950 Bedford O type dropside owned by Ray Brett, seen here in the driving seat.

A classic 1981 Bedford TK Tautliner box van is seen here in Somersham, with Jilly and Andrew Green of Beccles, heading for the rally site. This fine vehicle is still used every day for local deliveries in the potato business.

is owned by Keith Evans. During the course of the day people riding on the free vintage coach service had a great opportunity to photograph the coach outside the old garage, complete with an original AA call box - a unique experience indeed! I had the opportunity to chat with Howard Dolby, who was born and bred in Somersham, and who actually lived with his parents at this garage which they had owned since 1929. The garage actually had the first hand operated petrol pump to be installed in

Somersham, replaced in 1947 by more modern pumps of the 'Beckmeter' type.

One of the first lorries to pass through the village was Derek Styles' smart 1949 Bedford KD dropside, registered KTB 598, which spent most of its working life in the Manchester area.

Ray Brett, a haulage contractor from Wisbech, brought along his superb 1950 Bedford OLBC dropside, KNX 356, in the livery of Bretts Transport. The thing that puzzled me was this fine lorry in dark blue had in fact a red bonnet. Ray kindly explained that all the Bretts' lorries had a unique red bonnet during that period, a theme that was continued with all the commercials run by the father and son owned company.

Two fine recently restored Bedford pick-up trucks were brought to the event by Roy Woollard of Hemel Hempstead, on his hydraulic breakdown recovery vehicle. Firstly there was 1958 Bedford J1 TET 219, completely rebuilt literally from bits. Roy tells me that the cab was originally off a fire engine and came in bits, but now looks superb on this recent project which has taken 12 months to complete. His second restoration, a 1962 Bedford JO half-tonner, took only six weeks to complete from start to finish. Both these vehicles are a credit to Roy who certainly is a keen enthusiast and fine engineer. His present project is a 1952 O type tipper, we look forward to seeing that in the future.

I was intrigued to listen to Joe Ovel from Somersham, who brought his splendid 1936 Bedford BYC van DCD 567. This little 12cwt van spent the best part of its life delivering motorcycles and accumulators for wireless sets during the 1940s. Joe said that the van was on display at the Beaulieu Motor Museum, and having agreed the asking price with the owner it became Joe's pride and joy. This van has had nothing more than a good service and an MoT, which it passed with flying colours, and can now be seen around the town of Somersham.

Two very smart Bedfords were entered by the Kings Oak Nursery of Enfield, a 1950 Bedford "K" type with fruiterer's body and a unique 1952 Bedford/Scammell artic tractor unit registered SMU 452. The Scammell had two previous owners, and

Clive Matthews of Bishops Stortford is seen with his fine 1952 Bedford OSS artic unit and trailer.

A rare classic is this 1985 Bedford TL pantechnicon entered by Overseas Moving Ltd of Great Shelford.

A closer look at the Bedford belonging to George Mackay. George tells me that the Bedford will be restored, we look forward to seeing the finished job.

spent five years in the brewery industry and five years in brick haulage. It was eventually found derelict on a farm but bought and lovingly restored by the present owner Chris Norris. Both these vehicles looked superb and a credit to the owners who have painstakingly restored them to their present condition.

Owners of vehicles other than Bedfords turned out in support of this unique event. John Rignall brought his fine 1958 AEC Militant 6 x 6 flatbed 899 UKM, having recently completed the restoration. Local man John McGlade of Somersham, turned out with his smart little Thornycroft Nippy Star recovery truck, a vehicle seen at many events.

A superb 1977 Volvo F86 artic unit and trailer was an impressive sight, loaded up as it was with a 1950 Bedford O type flatbed, both owned by George Mackay who had driven all the way from Watford. Two fine ERFs were on display to celebrate this unique occasion. From the past there was a 1964 KV-cabbed CSC 235B two-axle rigid, and from the present the modern EC11. Both vehicles were provided by ERF Trucks Ltd of Sandbach.

My thanks to Richard Haughey, co-organiser of this splendid event, and to the owners of the vehicles for their assistance. ●

CECIL'S BEDFORD

By Bruce Cameron

No entry for the concours d'elegance here...

"I want you to do an article on a brown Bedford tipper - it's totally original, the chap lives down your way, here's his number, give him a call". These were the Editor's words, and I was passed a photograph and a magnifying glass - Peter had photographed the information board on the long wheelbase Bedford tipper at a recent rally. Thankfully, I could read the information...

Most enthusiasts will immediately cry wolf, saying Bedford never made a factory long wheelbase tipper, and I must say I think you're quite correct. Certainly this one never left the factory as a tipper, the biggest give-away being the identification plate under the bonnet - OLBD. This designation was given to model O (O), long wheelbase - 13ft 1in (L), 5 ton capacity (B), dropside truck (D). Perhaps more important to our Editor was not the fact it tipped or otherwise, but its originality - total originality.

The Bedford in question is a January 1947 5 ton dropside truck, believed new to Elliott's of Cheltenham who started out trading from Princes Street. I cannot tell you who in the company was a horse racing fan, but it is understood that all of the Elliott lorries were named after steeple chasers. John scraped a extra layer of paint off the headboard revealing the name 'Desert King', so this one was no exception. The signwriting on the doors was also covered by a quick coat of paint. This was removed uncovering the original and, as it turned out, a second version of the signwriting which had been redone when under the ownership of Elliott

Look carefully and you can just make out the lettering change from 'Transport Contractors' to 'Transport Contractors Cheltenham Limited'.

You can just see the end of the hydraulic oil release cable. Look carefully at the bearer channels on top of the main chassis rails, once again cut with gas.

Look at the cut end of the cross brace, once again cut by gas leaving a rather ragged finish.

The braces read 'Street is Neat' - John is also a fan of American vehicles - the Bedford connection is quite appropriate.

rothers. The original lettering read 'Elliott Brothers, Transport Contractors', this changed to 'Elliott Brothers, Transport Contractors Cheltenham Limited'. The second owner was a farmer, again from the Cheltenham area. At this point in its life, the Elliott signwriting was covered by previously mentioned brown paint and in 1963 the Bedford was converted to an underfloor tipper, and used for grain, stock and general haulage. Some tickets for coal also

suggested this to be one of its more regular loads. Looking at the general standard of workmanship and before verifying the history of the conversion, John felt that it was likely that the farmer himself had done the conversion (and he was later able to confirm this). The general engineering was quite good, but details such as ragged edges left by the cutting torch on the additional members necessary for the

tipping body left a little to be desired. The tipping gear is fitted well. Looking at other parts of the conversion, the rear springs (ten-leaf rear with six-leaf helper) also tell a story - the front spring hangers are riveted to the chassis whereas the rear hangers are bolted on and bear the 'Telehoist' name on the castings - the rear spring hanger also provides the hinge point for tipping gear.

The hydraulic pump for the tipping gear bolts on to the side of the gearbox and is engaged in the cab via a lever - you need to dip the clutch to engage the tip gear. To lower the body after tipping, a lever is pulled that connects to a valve via a wire cable, allowing the oil back into the reservoir.

While looking underneath, John pointed out a few bits of straw caught by the bell housing after the lorry's long stay in the barn.

Apart from the replacement of the body's wooden longitudinal members with 5 by 2.5in steel channel sections, used to strengthen the body and act as bearers, the only other easily visible sign of the conversion is a slight rise in the height of the body, which is completely original and unchanged, retaining all the strap work and retaining pins, etc.

Before John bought the Bedford it had been stored in a barn for many years, but had not suffered as a result. The only unusual feature I could spot was wood worm - not usually associated with a moving target and undoubtedly contracted during its long rest.

Very little work has been required so far to get her in running condition, at the time

Apart from replacing a few wires and cleaning the carburettor, the engine is untouched.

The cable that opens the valve allowing the hydraulic oil back into the reservoir can be seen running from the base of the ram to the cab.

Rear spring hanger - this also carries the pivot point for the body and would have been replaced as part of the conversion.

The body is original right down to the woodworm, no doubt gathered while stored in the barn for many years.

If you cannot make out the model designation, it is an OLBD, O series, L - long wheelbase, B - 5 ton capacity, D - dropside truck.

The front spring hanger is the standard item, unchanged in the conversion.

of writing although not yet 'on the road', it was up and running for the rally field. The brakes were completely overhauled with new seals, new master cylinder, etc. New tyres have been fitted - a matching set of 7.50x20 Michelin radials as no crossplies could be found (John commented that he's really going upmarket with radial tyres, and very lucky to find a matching set of 7). A little tuning to the engine, the removal of an alternator that had been fitted (replaced with a dynamo), new plugs, leads and points, and a good clean of the carburettor have also been necessary. A new silencer has been fitted too, but at the time of writing John still needs the front exhaust pipe.

Over the course of the years some features of old vehicles you almost expect to see updated with modern technology - on the Bedford the vacuum operated wipers and rather poor headlights are prime examples, yet even these are the originals as fitted from new.

Later on that day, we all popped back to John's for a cup of tea and John showed me his 1954 stepside Chevrolet 3100 pick-up - powered by a six-cylinder petrol engine that bore an uncanny resemblance to the engine of the Bedford. In fact, for the greater part these very dissimilar vehicles share the same engine, echoing their parental origins.

One chap tried to buy the Bedford from John - he didn't really want to sell it but advised the prospective purchaser that anything could be bought at the right price, but not until after the Welland show - the last commitment of the year. Thankfully, the buyer did not turn up. To be quite honest I don't think John will ever

part with the Bedford.

It is so pleasing to see vehicles in their original condition, as they were used and modified to meet the commercial pressures and requirements of the day, but it is indeed rare now to find a vehicle of such advanced years that does not require 'major surgery' to keep it in a safe and usable condition. Now and again an absolute gem does surface, and John's Bedford is no exception. Given the opportunity, this is the type of vehicle that would give many of us great pleasure to own, and even to use, for those that are not kept in concours condition can surely work a little for their keep.

On behalf of *Classic & Vintage Commercials* I would like to pass on sincere thanks to John, his wife Joan, and Les for their hospitality and assistance in putting this article together. ●

Although showing signs of its age, the cab of the Bedford is surprisingly tidy.

The following specifications are taken from the Instruction Book for Bedford 3, 4 and 5 ton chassis and Bedford Scammell Tractor, dated March 1948:

Engine
Type: six-cylinder petrol
Bore and Stroke:
3.3/8in (85.72mm.) x 4in (101.6mm)
Piston Displacement:
214.7cu. in (3519cc)
Max. Brake Horse Power:
72 at 3,000rpm
Max. Brake Torque:161ft lbs at 1,200rpm
Compression ratio: 6.22:1

Petrol System
Carburettor:
Zenith 30VIG-3 down draught.
Petrol Feed
By AC petrol pump driven by camshaft
Petrol tank capacity 16 gallons

Cooling System
Radiator and Cooling System:
Film-type copper core. Thermostat incorporated in cooling system. Impeller water pump with self-adjusting gland. Four bladed fan.
Air flow assisted by cowl.

Ignition
Type: Lucas coil and distributor with fully automatic centrifugal governor and vacuum advance controls.
Spark Plug: AC type VF9, 14mm thread. Gap .028in to .030in
Firing Order: 1-5-3-6-2-4

Electrical Equipment
Battery: Exide type 3XCZ-15M. 6 volt; 100amp-hour capacity at 20-hour rate.
Generator: Lucas model C.39 PV-L/O, 6 volt. Belt-driven. Compensated voltage control.
Starter Motor: Lucas model M45G-P29, 6 volt.

Transmission
Clutch: Single dry plate with flexible centre. Pedal adjustment to compensate for wear. Radial type clutch release bearing requiring no lubrication.
Gearbox:
Spur gears - 4 forward speeds and 1 reverse.
Ratios:

First	7.22:1
Second	3.47:1
Third	1.71:1
Top	Direct
Reverse	7.15:1

Rear Axle: Full-floating type with taper roller bearings located on the outer ends of the axle tubes to carry the load. Spiral bevel gears with pinion straddle mounted. Four-star type differential gear.
Rear Axle Ratios: 4.714:1, 5.286:1, 5.429:1 or 6.2:1

Front Axle and Steering
Front Axle: Heat treated drop forging of "I" section beam. Stub axle load taken by plain-type thrust bearings. Taper roller hub bearings.
Steering Gear: Semi-irreversible worm and wheel. Self-adjusting ball joints to tie-rod and steering connecting rod.
Turning Circle Diameter: 59ft (17.98m)

Frame
Type: Pressed steel channel section

Brakes
Type: Lockheed hydraulic, operating on all four wheels. Operation assistance by Clayton-Dewandre vacuum servo. Handbrake operates rear brakes mechanically.
Size of drums: Front - 13in diameter.
Rear - 14in diameter.

Wheels and Tyres
Wheel Size: 5.00 flat base x 20-4.9in off-set rims all round.
Tyres: 34 x 7 Heavy Duty, all round.

Bruce Cameron reports on the restoration of three ex-Jersey trucks, one of which had a most unusual conversion for use on the Channel Islands. By coincidence, all are now actively preserved near Bruce's West Country home!

Jersey Roy

Kevin Patch with Bedford and his boy, Jake.

Coincidence is often the impetus for me to put pen to paper or, to be more precise, finger to keyboard. This was the case when I realised that I had 2 Bedford MSD's and an Austin K2 right on my doorstep, all of which had been imported from Jersey, where they'd apparently carried out similar duties during their working lives.

Who could resist finding out more?

Kevin Patch
1950 Bedford MSD

Originally shipped to and registered in St Helier, Jersey (the supplying garage was Cole's), Kevin's MSD is a 1950 model. Researching the history of this vehicle, it was discovered that the first recorded keeper was Mr Clifford George Pallot, a

Period British registration number hides Jersey history.

Immaculately restored Bedford previously had 'loft' fitted over the cab and a longer tailgate, allowing livestock to be carried.

Engine was clogged with silt when the Bedford was purchased, hence it required major flushing out!

farmer from St Martins. The lorry was used as a general purpose vehicle and carried potatoes and other vegetables grown on the farm. Certainly the most significant task for many of these lorries was carting the Jersey Royals to town, as this formed a significant part of the island's economy. In the summer time, those that did not have horse transport, would take the lorries into the fields to move the crops - cereal and roots for example. Clifford Pallot was also a cattle breeder, and had the lorry fitted with a longer tailgate, allowing livestock to be carried. The tailgate was quite short necessitating a loading ramp to be used in conjunction with it. Kevin, a steel work designer/fabricator by trade, found traces of shotblast grit in the cab which led him to believe that the body had been rebuilt. The previous owners also re-wired it.

The MSD was effectively a non-runner, suffering numerous overheating problems, when Kevin, Chairman of the National Vintage Tractor & Engine Club's North Somerset Group, bought it. The radiator was removed and flushed, dislodging a large amount of silt. Seeing this, a core plug from the engine was removed to find the same story - the engine was also full of silt. The block was flushed several times, using acid solution. Now in its sixth year of rallying, the little Bedford shows no sign of its past troubles.

Now, a 'cattle float' previously fitted has gone, as has a loft fitted over the cab that was still there when Kevin became the MSD's proud owner. This attachment was in very poor condition, so was removed. Not too much of this Bedford's history is known after it left Jersey, though there are details of two owners.

Kevin is keen on rallying the Bedford and each year takes part in the Glastonbury to Burnham run and Bournemouth to Bath runs, among others.

I have had the pleasure of several trips out in this lorry. It has its plus points and minus ones. On the plus side, Kevin has fitted a 'faster' differential from a Bedford TK, making for some embarrassed would-be overtakers, and on the negative side, I can advise from experience that the demisters (non existent) and windscreen wipers are not up to much! The wipers are vacuum operated - they slow down, and all of a sudden they burst into life. As I commented while making tape recorded notes during last year's Glastonbury to Burnham on Sea run: "Kevin fiddled gallantly with the vacuum control of the wiper all evening but to no avail, the single blade refusing to move from its resting position, pointing vertically down, stuck in the middle of the screen."

Still, it's all part of the fun!

Denis Stone
1947 Bedford MSD /
1947 Austin K2

Having given up a long standing relationship with the sea, Denis Stone, by then a semi-retired driving instructor, was at a loose end. Flicking through a magazine he saw two lorries advertised in St Martin's, Jersey - a Bedford MSD and an Austin K2. Luck must have been smiling as not only was he off to visit Jersey the following week, but both lorries were as yet unsold.

Collected from St Helier, he was taken to a farm, where he discovered the vehicles were still quite literally on the road and being used - he had a brief wait until the lorries returned from the fields. Both were in fairly poor condition. The Bedford had its petrol tank tied on with string, the wings were flapping, and one of the rear wheels did not touch the ground at all. The Austin was in a similar state of disrepair.

Transport arrangements were made and Denis returned home without lorries, the cost being £55 per lorry unaccompanied on the ferry. Once home, in the cold light of day they certainly looked rather worse than he had remembered.

The Bedford was the first to be tackled, stripped to the bare chassis, this was grit blasted. Considering that Jersey is a small salty island, there was very little rust in the bodywork, the main repairs required around the passenger side window. Methodically all parts were removed, grit blasted and painted. Interestingly the lorry came with a spare pair of rear wings, later purchased by Kevin for his MSD. Very few spares were required to complete restoration. However, the restoration took four-and-a-half years, this was done mainly in the open or in a small domestic garage. Denis commented that the paint on the Bedford was particularly difficult to remove, and this probably contributed in no small part to the lorry's good state of repair when purchased. Steve Trott, of Chard Autobodies, completed the paintwork,

No, it's not a factory conversion - the chassis shortening was a skilled job carried out on Jersey.

Original 3.0-litre Austin engine has been replaced with 4.0-litre unit

leaving the seats to be re-upholstered and headlining replaced. The Bedford has now been on the road for several years, and has been entered in the Bournemouth to Bath run each year since, and a number of local events.

Denis thinks the lorry's great to drive but with terrible lights - as he terms it "two good quality candles". The wipers are also prone to working occasionally when you slow down (something to do with vacuum operation...).

Moving on to the Austin, this received similar treatment, with many of the chromium parts re-plated, new shackle pins and bushes fitted and so forth. Skipping the restoration detail, this lorry has perhaps the most interesting story to tell, as it is a short wheelbase, and yes, you're quite correct - there were no factory examples.

Now for another piece of luck.

Ex-*CVC* editor Peter Love suggested I have a chat with Lyndon Pallot, whose family had at one time quite a substantial trade on Jersey in shortening lorries. As Peter predicted, the Pallots had converted this vehicle, removing on inspection, 2ft 3in to be precise. Jersey being a small island with rather narrow twisty lanes did not lend itself to a long wheelbase, there were many small

farms, most of which had large granite gateposts, designed for the horse and cart - this necessitated the use of, by today's standards, a small vehicle.

Around 120 conversions took place, with most makes included, including a few Morris Commercials, but Austins numbered some 85 per cent of the total. As Lyndon commented "You could not buy the product, so you had to adapt". Bedfords coming in both long and short wheelbase variants did not require such modifications.

Unbelievably, the Pallots had also lengthened a few ex-military American Dodge trucks - these were rather too short, so these had sections added in (utilising the pieces removed from the Austins, no doubt!).

Lyndon vividly recalls how the shortening process was carried out. The first task was to remove the bodies with a block and tackle, then drain off the oil, fuel and water, removing the battery. Next the chassis channels were marked and cut with a hacksaw. A flexible grinder came next (mounted on four small little castors - made by 'Biax'), and this was used to 'V' the chassis members to lay the weld in. The top and bottom faces of the chassis channel sections were welded first. Next, using a block and tackle, the lorry was turned on to its side. "The frightening part was getting it high enough," Lyndon said. The inside of the (now) lower chassis channel section and outside of the (top) chassis section were then welded. The lorry was then tipped the other way, in order to weld the other side in a similar manner. This practice was used for the first 50 to 60 vehicles. After this, technical advances to welding consumables

Denis Stone with his MSD.

A Like the Austin, the Bedford bears signwriting of previous Channel Islands owner.

Bedfords were wont to foul their spark plugs on short Channel Island journeys.

and increased skill, allowed the welding procedure to be completed in situ, without the use of the block and tackle. This saved some considerable time in the conversion process.

Propshafts were shortened, using the lathe to ensure that the universal joint was re-welded completely square to the shaft. Brake pipes and the handbrake rod were shortened.

Focus then turned to the bodies. It was found that with the configuration of the body crossmembers the most practical way to approach the work was to turn the body round, so what was previously the back was now closest to the cab.

Once shortened, the trucks were repainted, lined and signwritten for those that required it. Some trucks were supplied to other dealers, who may have done their own painting.

Restoration of this Austin was completed in the first half of 1999. Says Denis: "The Austin in comparison to the Bedford is perhaps of higher quality, fitted with such luxuries as electric windscreen wipers and wind-up windows."

One last snippet of information recalled by Lydon is that Bedfords were rather prone to oiling their plugs while used on the short island journeys - to avoid this, an extension tube was screwed into the spark plug hole, moving the spark plug out around 1½ inches, this seemed to cure the problem in most cases, whilst in others, rebores were not uncommon after just 10,000 miles, Denis's lorry has had an engine swap during its time on Jersey, which would support this fact.

If you would like to see these lorries out on the road, why not go along to the

Glastonbury to Burnham on Sea road run on June 3, in aid of Children's Hospice South West. An evening kick-off at Glastonbury, ice creams at the mid-point of Burnham, and then back to Glastonbury for a buffet. A good day out!

⬤

*CVC would like to thank to Kevin Patch, Denis Stone and Lyndon Pallot for their help with this article.

VEHICLE SPECIFICATION

Bedford MSD
Registration: J10109
Date of First Registration:
5th February 1948
Colour:
Grey/Red (later changed to blue)
Chassis Number: 70816
Engine Number: KM79264 (KM30026 fitted as replacement)
Seating Capacity: Two
Recorded Keepers:
1st Keeper (5th February 1948)
Mr A.P. LeMaistre, Peacok Farm, Trinity.
2nd Keeper (Date unknown)
Mr G. LeMarquand, Le Catillon, St Mary's.
It is understood that Mr Gordon Philip LeMarquand and Mr A.P. LeMaistre swapped vehicles - Mr Gordon Philip LeMarquand had a military type Bedford, registration J6557 - he kept this registration and put it on Mr A.P. LeMaistre's Bedford.
3rd Keeper (18th May 1989) Mr Denis Stone, Glastonbury, Somerset.

Austin K2
Registration: J4136
Date of First Registration:
2nd November 1947
Colour: Light Green / Dark Green (later changed to blue)
Chassis Number: 113905
Engine Number: 1K-128019 - 3,459cc
(1K-273169 - 4.0-Litre unit fitted as replacement)
Seating Capacity: Two

Recorded Keepers:
1st Keeper (22nd November 1947)
Philip LeMarquand, Prospering Farm, St Mary's
2nd Keeper (9th January 1963)
Edward James LeMarquand, LaTourelle, St Martins
3rd Keeper (23rd May 1973) Gordon Philip Le Marquand, Le Catillon, St Martins
4th Owner (18th May 1989) Mr Denis Stone, Glastonbury, Somerset

Bedford MSD
Registration: J5686
Date of First Registration:
16th March 1951
Colour: Maroon
Chassis Number: MSD209222
Engine Number: KM209118
Engine Capacity: 3,519 cc (28 hp)
Unladen Weight: 39 cwt
Seating Capacity: Two

Recorded Keepers:
1st keeper (16th March 1951)   Clifford George Pallot, La Poudretterie, St Martins, Jersey.
2nd keeper (1964) Clifford and John Pallot, La Poudretterie, St Martins, Jersey.
3rd keeper (26th November 1975) Clifford George Pallot, Les Jardins, Fliquet, St Martins, Jersey.
4th keeper (Date unknown) Billy Bowens, Sandy Lane, Chobham, Surrey.

The Archers of Northallerton

David Reed returns to his home town to meet father and son team Ken and Paul Archer, whose 1956 Bedford A5LCG is a well-known rally star. They reveal two further restoration projects... called Cecil and Josephine!

Photos: David Reed.

Your home town always provides fond memories, so Northallerton in North Yorkshire is a special place for me.

Fans of heavy haulage will still associate the town with the long-gone firm of Sunters, and I can well remember their mighty Scammells manoeuvring the heavy loads around the streets. Thankfully very much still in operation is removals firm Archer's, and it was the chance to meet father and son Ken and Paul Archer, and their collection of restoration projects, which prompted my trip 'home.'

Their collection includes the well-known 1956 Bedford A5LCG, NDN 618, featured on the front cover of September 1997's *Classic and Vintage Commercials* while taking part in the Tyne-Tees Run.

This vehicle was supplied new by Leedham's of York to George Kilvington, a farmer from Easingwold. No heater is fitted - that luxury would have cost an

A well-known sight on the rally circuit is the Archers' wonderfully restored Bedford A5LCG. The 'load' contains living accommodation used at events, and known as 'the office."

Bedford O-Type proved a temptation too good to miss, though it's a major project.

Ken Archer contemplates work needed on the Bedford.

completed by the late John Stevenson, from Northallerton.

The Bedford has been a regular rally attendee ever since, carrying a shed on the back known as 'the office.'

Fine though this Bedford is, it wasn't the main reason for my visit. I'd heard a new project was underway, a 1956 Atkinson L1586 fitted with a Gardner 6LW engine, registration NVP 63 and chassis number FC 2543.

Built for the British Transport Commission in 1953 and first registered on January 5, 1954, the Atkinson was transferred to British Road Service, being based in Birmingham for its entire working life with this operator, and receiving several different fleet numbers.

In April 1966 it was withdrawn and sold to Cecil Shipley, a showman based at Richmond, North Yorkshire.

He was delighted to discover that a reconditioned engine had been fitted, and a brass plate indicated that it had been overhauled at the BRS workshops in York. Mr Shipley also replaced the original flatbed with a Luton box van body from an ex-Army six-wheel Leyland, partitioned off for the showman's living accommodation.

The spare wheel carrier was removed to make way for a dynamo to provide power for the fairground rides. Being slung under the chassis at the back, the drive was arranged via a propshaft mounted on top of the crossmembers under the floor. The axle driveshafts were unbolted when power was needed and the dynamo shaft bolted direct to the gearbox flange, the dynamo being driven through the gearbox.

Later on, it was decided that as the Atkinson was due to be plated and tested, and two of the tyres were completely worn out, it would be better to remove the second steering axle to save the purchase of the two tyres and help the manoeuvrability of the lorry when working in the showfield. Many eight-wheelers were thus converted by showmen over the years.

In the 1970s, the Shipleys operated their vehicles from a site at Flamingo Park Zoo, and the Atkinson was parked up on a caravan site at Stainforth, near Doncaster, its last test and plating certificate running out on January 31 1973.

Time passed. Ken and Paul had heard that there was an Albion Caledonian reported to be lying on a scrapyard in the Doncaster area. "We got there and the owner revealed it had been cut up, which was a great shame, but he did tell us that there was this Atkinson six-wheeler he

extra £15. After covering a mere 42,000 miles in farm service, the Bedford was bought by Ken and Paul from Mr Kilvington in December 1990, and named George in honour of the first owner.

"It had never spent a night outside," recalled Ken, though work on the cab was needed. The engine, maybe not surprisingly, didn't need any attention, though has since had new pistons after dropping a valve. A repaint was

Somewhere in there lurks a Perkins P6!

Cecil the Atkinson L1586 has benefited from a stripdown to the chassis work on the cab framing. He'll hopefully be on the road next year.

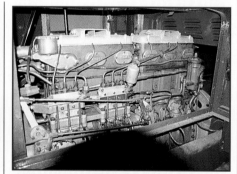

Atki's Gardner 6LW was reconditioned in BRS days.

knew of, which might be for sale.

"Mr Shipley was approached, and initially Paul was put off by the vehicle only being a six wheeler. When we were told it was originally an eight wheeler he pricked his ears up," said Ken.

The lorry, now named Cecil, like George the Bedford in honour of its previous owner, was purchased and a working day set aside to prepare it for removal. After bleeding the fuel system and fitting two 12 volt batterys, the engine fired up almost straight away.

Next, the wheels were jacked up one at a time to ensure they were free and the lorry was mobile. Mr Shipley thought this was a waste of time as he had made sure that the brakes were well backed off when he'd parked the lorry ten years previously! "To everyone's surprise he was right - all of the wheels turned," recalled Ken.

Next, the fitted dynamo and propshaft were removed from the chassis and winched away.

Recovery was carried out on a lowloader, and one of the other occupants on the site, watching the loading, thought he had an axle to replace the one which had been removed.

A check with Kirstall Axles confirmed he was right, and the axle was bought, together with springs which could be altered to fit the Atkinson. Another

example was found in Lincoln to provide the correct hubs, and another Gardner 6LW engine has more recently been bought to act as a spare.

Since then, Cecil has been stripped down (the lorry, not the vendor!) to the chassis frame for restoration, this being sandblasted and painted. The famous Rush Green Motors provided parts to

restore the steering assembly.

The fitted van body was found to be past repair, and so Cecil is being restored to flatbed specification. The front panel has been repaired where necessary, with a mould having been made to restore the sidelamp assembly.

A third vehicle has joined the restoration fleet, however, a Bedford 0-Type tipper with a Perkins P6 diesel engine. This vehicle worked for Ivor Govier and Son of Blagdon Hill, Taunton, Somerset, and was sold after many years to enthusiast Arthur Metcalfe, from whom the Archers bought it.

"The Bedford, christened Josephine after an inscription on a cab door, just seemed a project we wanted to do, though really we bought it because it was there!", admits Ken. "A new cab is needed, the engine is seized and there's a lot of work to do, but we'll get there eventually."

So far, the Bedford's registration number is a mystery. There having been no plates with the vehicle, the Archers therefore assume the registration number has been transferred to another vehicle. They would welcome any information on this.

Also in the fleet is KVN 729F, a Bedford CA with rare Martin Walter Luton style van conversion. Owned by the family from new, it's painted in the livery of subsidiary operation Richard Thorpe of Ripon.

Archer's was founded by Ken's grandfather, Percy William Archer in 1920, originally selling Garner tractors before moving into removals and storage.

Ken started restoring stationary engines 20 years ago, and even obtained a Garner, passing on to commercial vehicles and Austin Somerset and Devon cars.

Although both Ken and Paul are very much involved with the haulage industry and spend their weekends working on the restoration projects, they never lose their enthusiasm, Ken saying: "We never get tired of trucks. They're in our blood." ●

Atkinson as discovered, the box van body being well past redemption. It is being restored to original flatbed spec!

Bedford

AT WORK

BEDFORD has a special place in commercial vehicle history. What started off in 1931 as an Anglicised adaptation of an American Chevrolet design soon became a world famous marque in its own right. The success of the early Bedfords was largely down to their rugged build, their smoothly powered six-cylinder petrol engine and their very competitive pricing. Thus many transport contractors began with one of Bedford's low cost workhorses. Their profit earning capabilities put them in a class apart. Many hauliers who started in a small way eventually worked their way up to big fleets, running maximum weight trucks from the likes of Leyland, AEC and Foden but it was the Bedford that set them on course to profitability.

Peter Davies joined Vauxhall Motors, the manufacturers of Bedford trucks, as an apprentice in 1957 and served 29 years with the company's marketing department. As a transport enthusiast he took every opportunity to rescue old photographs and brochures when they came up for disposal, and over the years he built up a fascinating archive. In this special supplement Peter digs into his collection to bring us a unique record of Bedford's most famous models through the years. In addition to the historic photographs, the supplement charts the diverse styles of sales brochures over the years - reflecting the different tastes of each era.

This superb shot of a 'WLG' 2-ton dropside truck was taken in 1937 at Jessup's Lime Products of Polhill, Kent. The two towers in the background are not oast houses as one might assume, but Jessup's lime kilns - a notable historic feature at the 700-year-old site. The truck was operated by haulage contractors Mayes Bros Ltd.

Manchester's Smithfield Fruit Market is the setting for this busy scene depicting two hard-worked 'WLG' two-tonners. The one on the left dates from 1933 while its stablemate is a 1931 model from the first year of production. The lazytongs mirror arm on VU7922 is a novel feature. The shot was taken in 1935 and both trucks carry Bedford Driver's Club badges. The club was formed in 1934 and one year later boasted over 20,000 members.

This 1932 'WLG' was one of a fleet of 34 similar vehicles purchased by the Southern Railway Company and is seen at the Bricklayer's Arms Goods Depot in South London. The coachbuilt cab and bodywork were built by Spurlings Ltd of Hendon. The sideguards are an interesting feature - such attachments did not become a legal requirement in the UK until 50 years later.

A far cry from today's refuse collection vehicles is this 1932 'WHG' dustcart photographed in Sussex. It was new to Burgess Hill Urban District Council Cleansing Dept. Loading it must have been backbreaking work. The lift-up hatches are operated by hand levers through a somewhat 'Heath Robinson' style linkage. The refuse cover was removable so that the truck could be used for general duties as well.

Above: Giles & Bullen of Kings Lynn in Norfolk certainly got their money's worth out of their early Bedfords if the 1933 'WHG' artic is anything to go by. Such machines were joint products with firms like Carrimore, Truck & Tractor Appliance (forerunner of BTC), Flexion and Muir Hill. Bedford never quoted payload ratings for such conversions but they nominally doubled the standard capacity. On the left, at their Norwich Depot, is a standard 2-tonner while on the right is a 1935 'WTH' artic advertising a Birmingham to Eastern Counties daily trunk service.

Right: Dobbin probably took a dim view of his owner's decision to purchase this brand new 1937 'WS' coal lorry. Perhaps his blinkers eased some of the trauma! The new acquisition of Coal Merchant W.F. Holt of Ponders End, Middlesex, poses outside Arlington Motors, one of Bedford's largest dealerships.

Could the registration number of this 1935 'WS' 30 cwt van be a coincidence or was it an early instance of reserving a 'cherished' number? The vehicle was based at Oxo's Leeds depot and 'CUB318' was conveniently issued at the time when delivery was taken. On the side of the van is proudly displayed the Royal Warrant of King George V.

On a scale of one to ten this 'WTL' surely scores a ten. In typical Scottish style it is beautifully signwritten and lined out. It was new to C. Robertson & Co of Springwells, Blantyre in Lanarkshire in 1936. Evidently no expense was spared

This excellent shot taken near the outskirts of Coventry in 1938 captures two examples of Bedford's best loved classic - the 'WTH' 3-ton tipper. The '3-ton' rating was misleading since Vauxhall Motors themselves readily advertised it as 'the truck for a 50% overload'. That made them 4¹/₂-tonners, but many operators treated them like 6 to 7-tonners and they still soldiered on! These are loading up granite chippings from a roadside stockpile. On the left is a 1937 vehicle while its sister is a 1938 model with the updated 'bullnose' grille. They belonged to Watkin & Grew of Coleshill. The impostor hiding its face in the background is a Fordson V8 'Model 51'.

Above: George Dady of Anerley in South East London formed a small sand and gravel business in 1935 with just two tippers. Within three years he had 18 such vehicles - all Bedford 'WT' 3-tonners. This 1938 'WTH' is a typical example of Dady's pre-war fleet which was once a familiar sight on major developments throughout the London area.

A **CLASSIC** and Vintage **COMMERCIALS** supplement

...on this magnificent machine - it even has the optional chromium plated radiator surround which did not come cheap at £2 10s (£2.50). Quite a tidy sum when you consider that the chassis cab complete was only £293.

It's just as well this brave guy had a head for heights - or maybe he just lost the toss and was made to overcome his vertigo in order to pose in this shot. This 'WLG' 2-tonner with coachbuilt crew cab and cable operated tower wagon for street light servicing was new to Lanarkshire County Council, Bellshill District, in March 1938.

Above: Subtle might not be the first word that comes to mind when describing this 1938 'WHG' advertising van created to the design of the Michelin Tyre Company of Stoke-on-Trent. It highlights the road-holding virtues of a new tyre called the 'Stop'. The front wheel appears to be fitted with the advertised product.

COMPL

BED

VANS, TRU

M

VAUXHALL

You see th

RANGE

ORD

& BUSES

BY

TORS LTD.

verywhere

Above: Bedford chassis were a popular choice for a whole range of fire appliances. This 'WLG' based pump escape for the Henley on Thames fire brigade is being demonstrated at the Vauxhall factory in 1938 prior to delivery to the customer.

Left: During World War Two most of Bedford's production was turned over to military vehicles and Churchill tanks. Approximately 250,000 military trucks were produced including a high proportion of MW '15 cwt' models. Many were specially waterproof like this example in full combat trim disembarking from a landing craft.

Below: Austerity pattern front end sheet metal gave wartime civilian Bedfords a completely different appearance. The wide bonnet was standardised to accommodate civilian and military cabs, the latter in some cases requiring a larger than standard air filter for desert use. 'O types became 'OWs' and all were built as 5-tonners with helper springs and 34x7 rear tyres. In artic form they were rated as 8-tonners as with this example in the fleet of Plymouth Transport Co Ltd.

This OST 'Code 40' 5-ton tipper dates from 1939 and was eleven years old when this shot was taken at John Williams Transport's premises in Cardiff. Standing next to the Bedford is a 1918 Burford truck which has been converted to pneumatic tyres and was still in use on internal duties in the 1950s.

Above: Hard at work laying tarmac in Glasgow's Milton housing estate is this 1946 'OSBT'. The tipper belonged to Carntyne Transport Co and is seen working in conjunction with a Barber Greene. Rollers on the Barber Greene contact the rear wheels and push the tipper along while tarmac is tipped into the machine. The picture was taken in 1952.

Above: The 1950 Earls Court Show saw the public debut of the new 'Big Bedford' for seven-ton payload. The pleasing lines of its completely new forward control cab made it stand out as one of the most talked about trucks at the show. Standard engine was Bedford's 110bhp 4.9-litre petrol six. A diesel option, the Perkins R6, became available in 1953. The petrol 'S' type was half the price of a Leyland Comet and a third of the price of a Maudslay Mogul yet carried the same payload. This 'SSTG' with standard 6 cu yd end tipping body is seen loading hot ash at a power station in Edinburgh.

Left: Perhaps one of the most memorable Bedford fleets was that of David Jones & Co Ltd, the large wholesale grocery firm based at 33 Redcross Street, Liverpool. A huge tonnage of imported foodstuffs coming into Liverpool Docks was distributed to retail shops all over the UK by their fleet of 62 bright yellow trucks. David Jones' speciality was importing, blending and packing 'Golden Stream' tea which was delivered to approximately 10,000 shops throughout England and Wales.

Above: This evocative scene in South East London graphically recalls the early fifties, before the days of 'double yellows'. Sid Perrett Ltd, the owners of this 1951 'SLCG', were based at Neate Street in Camberwell. It is loaded with crushed bone which went into making fertiliser and glue. To cope with frost in the winter months the truck has an electric screen demister. One wiper and one rear view mirror were deemed sufficient in those days.

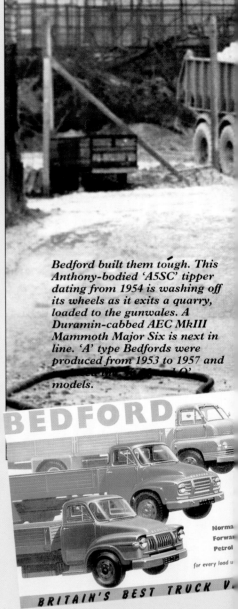

Bedford built them tough. This Anthony-bodied 'A5SC' tipper dating from 1954 is washing off its wheels as it exits a quarry, loaded to the gunwales. A Duramin-cabbed AEC MkIII Mammoth Major Six is next in line. 'A' type Bedfords were produced from 1953 to 1957 and ———————— 'D' models.

Left: What schoolboy could not have wanted to be let loose at Meccano's famous Binns Road factory in Liverpool, surrounded by those fabulous Dinky Toys and Supertoys! This 'A' type has the 'Jekta' moving-floor van body which Dinky reproduced in miniature on the 'A' type's successor, the 'D' type of 1957. The truck seen here is loading pallets of new castings at Dinky's Speke plant on the outskirts of Liverpool, to be taken to Binns Road for painting and assembly.

Facts about the NEW BIG BEDFORD

Left: 1952 'SAC' 10-ton articulated bulk liquid tanker was part of a large fleet operated by Pickfords' Tank Haulage division based at Marshgate Lane, Stratford in East London. The 'S' type tractor was on an ultra short wheelbase of 86in and most were built for operation with automatic coupling semi-trailers as in this case. The picture dates from 1957 and the unit obviously had five years' hard graft to its credit.

Left: The 'RLC' earned a reputation for its 'go-anywhere' capability. Based on the 'S' type, it was the first 4x4 truck to appear after World War 2 during which the famous 'QL' was produced in large numbers for the armed forces. Some of the QL's pedigree rubbed off on the 'RL', which became the leading general service cargo truck for the British Army as well as selling in thousands to other military customers throughout the world. It was rated as a 3-tonner in military guise. This early civilian example is equipped with a lime-spreader body by Atkinsons of Clitheroe and was based in the Barrow-in-Furness area.

Left: Full of atmosphere, this scene in Liverpool Docks shows a 'KGA8' articulated tanker from the United Molasses Company, whose offices were in Blenheim Street. The steam shunting loco in the foreground is one of the veterans of the Mersey Docks & Harbour Board's railway which ran the whole length of the dock road. The 'KGA8' was introduced in 1960 as the heaviest model in the new 'TK' range and grossed 17.2 tons gcw. Initially it was powered by Leyland's O.350 'Comet' diesel and had Bedford's own close ratio 4-speed gearbox as standard. Usually the optional 5-speed box was specified. In 1963 it was uprated with a choice of Leyland's O.370 or O.400 engine, the latter having an 18.3 ton gcw and a 5-speed gearbox as standard.

A **CLASSIC** and Vintage **COMMERCIALS** supplement

Millers Bridge, Liverpool is the setting for this impressive shot of a 'J4A' artic loaded high with sawn timber. In the distance is Brocklebank Dock when Liverpool was still the country's major Atlantic port and the whole dock area was a hive of activity. The shot dates from 1960, just two years after the 'TJ' was launched. It turned out to be the longest running Bedford model and survived through to the 1990s even after the company was sold to AWD and subsequently to Marshalls of Cambridge. By that time all 'TJs' were exported in CKD form for overseas assembly. The 'J4A' itself was dropped from the range in 1966.

Another busy scene in Merseyside, this time at the Birkenhead end of the Mersey Tunnel. The distinctive tunnel architecture forms a dramatic backdrop to three Bedfords led by a 'KDL' loaded with 45-gallon oil drums from Frederick Braby of Liverpool. The 'KG' tipper behind has a load of scrap metal while the 'J3' van is one of a large fleet once operated by Droylsden-based Robertsons Jams. Just visible in the foreground is one of the Mersey Tunnel Joint Committee's Land Rover patrol vehicles.

Above: This 1981 shot is particularly significant in that it shows a new 'TL' and, in the background, a 'TK' which formed the basis for the 'TL'. The 'TK' was showing its age after 20 years and Bedford urgently needed a new design to stay competitive. Budget restraints ruled out a completely new range and the 'TL' was basically an improved 'TK' with tilt cab. Sadly, Bedford's fortunes declined rapidly during the eighties as they lost out on their traditional markets both at home and abroad. A bid by GM to strengthen its UK presence by the acquisition of Leyland Truck & Bus was blocked by the UK government and shortly after GM decided to pull out of UK truck manufacture completely. The news sent shock waves through the whole industry since Bedford had always been seen as a market leader.

What a superbly atmospheric photograph this is! A 'TK' artic of Alexander Scott (Contractors) Ltd, part of the J & A Smith of Maddiston Group, pulls up in the early hours at the Watling Street café near Friars Wash on the A5. The driver dutifully checks his ropes before going into the warm inviting interior for a well-earned 'cuppa'. The 'TK' was engaged on the firm's Glasgow-London trunk and was part of a 360-strong fleet which included quite a number of Bedfords.

The 'TM' was launched in 1974 and its early shortcomings were not resolved until an extensively re-engineered version came on the market in 1980. By then it was too late to restore the company's reputation in the heavy weight field. The last 'TMs' were first-class machines but before they could prove themselves Bedford closed down. This picture shows the very first 'TM', a 6V-71 engined 'TM3250' (officially an EWV8QDO), to enter service with a customer. It went to Frederick Ray Ltd of Leighton Buzzard.

Dating from 1932 this Bedford 'WS' 30 cwt truck of the Camberwell Borough Engineer's Dept is seen against a fascinating period setting. An AEC 'NS' double-decker, gas lamps and a host of advertising posters and shop signs all add atmosphere to this South East London scene from a bygone age. Note the Bedford's starting handle and its quaint bulb horn mounted on the bottom right-hand corner of the windscreen.

A **CLASSIC** and Vintage **COMMERCIALS** supplement

BEDFORD RALLY *Blooms*

Twice the size of the 1999 event, the Bedford Gathering 2000 proved a fine tribute to the marque. **Nick Larkin reports**

(Photos: Nick Larkin)

their cars benefited from a regular connecting service to St Ives Bus Station, using mainly Bedford buses.

Fifty years of the S-type lorry and SB bus/coach were celebrated. No-one ever intended the event to be pure Bedford, and many other marques were represented. As a result, ERF, which now owns the Bedford Genuine Parts Business, was there.

An excellent atmosphere, sales stands and other attractions all combined to make a great day out, with rain holding off until the last vehicles were leaving.

Next year's event, already scheduled for August 26, will celebrate 70 years of Bedford vehicles. See you there!

RESULTS
Best SB: Duple-bodied PPH 698, Geoff Heels; Best In Service Bedford: Duple-bodied YMT, YMJ 555S of Lodge's Coaches; Best Bedford Commercial:

You saw them everywhere, and here are just a few of the Bedford commercials and PSVs, plus a splattering of other makes, at Somersham. Seen in the centre of the line-up are two prize winners, RC Jeffrey's WLG commercial, BWP 934, which was voted best privately-owned vehicle. The event had doubled in size from last year, and made a great day out.

You saw them everywhere! It's impossible not to play on Bedford's old advertising slogan when describing a superb event paying tribute to the marque.

Not only did the Bedford Gathering 2000 (the third of its type) double in size from last year, but there were also many other pleasant surprises. Everything from an M-series fish-and-chip van to a CA caravanette was among the 150 Bedfords from across the country attending the event.

Sponsored by *Classic and Vintage Commercials'* owners Kelsey Publishing, the event included many vehicles still earning their keep, along with preserved examples. The show was organised by stalwarts of the Cambridge Omnibus Society and held at Dews Coaches' premises in Somersham, Cambridgeshire.

People not bringing

Ex-military RLs attract attention: MFF 917 is a 1953 4x4 and ESU 581, with a container body, dates from 1964.

Wychwood Brewery's OLBC, GSJ 761; Best Vehicle on Michelin Tyres: Lodge's YMT, YMJ 555S; Volvo Award: 1959 Bedford S-type Artic Unit, KEB 519; Cambridge Omnibus Society Cup: Kenzie's of Shepreth J2, KNK 373H; ERF Award for Best Working Bedford: TB Green's TK, KYW 328X; Best Non-Bedford PSV: John's Tours, A860 WAV; Best Non-Bedford Commercial: J Reynold's AEC Mercury, CHA 637K; Best Bedford Owned by Individual: RC Jeffrey's WLG commercial, BWP 934.

Big celebrations for the Bedford S-type, 50 years old this year.

You can just smell the aroma of beef-dripping-drenched chips mixed with soggy newsprint! This wonderful 1939 Bedford M-type mobile fish-and-chip shop belongs to restorer Eddie Doig of Wirral, Merseyside. DBM 224 spent its working life in the King's Lynn area of Norfolk, and straight after the rally, Eddie was due to take the Bedford back to its old haunts and reunite it with original chippie staff. The tour then continued to the Bedford's birthplace. Eddie is writing us an account of these events for a future issue of Classic and Vintage Commercials.

K-Type collection. On the right is Ray Hammond's restored 1939 dropside truck, which was recently re-united with its original owners, Sharp and Fisher of Cheltenham, Gloucestershire, which used it extensively during the war. The second K actually dates from 10 years later and is owned by Derek Styles from Fenstanton, Cambs.

Contrasting vehicles, but both loved by owner Joe Oval, who lives in the gathering's home village of Somersham. DCD 567 is a rare 1936 Bedford BYC van, new-to a Brighton cycle-shop owner and one of nine examples known to exist. The vehicle was stored at the National Motor Museum, Beaulieu, for some years before Joe acquired it. In contrast, SEW 315K is a 1972 CF pick-up, which although used by a local farmer for many years, has still only travelled 38,000 miles. "It would have been a shame to have seen it scrapped," says Joe.

Pre-war pair: ASJ 591 is a 1939 MLZ fire-engine, and FW is a beautifully-restored 1935 WLG.

Not only Bedfords starred at the event. A worthy winner of the Best Non-Bedford Commercial award was this AEC Mercury Artic of J Reynolds, a regular event participant.

Kelsey publishing director and Car Mechanics editor Peter Simpson (left) presents the Kelsey Publishing trophy for the Best Bedford Commercial to David Russell of Wychwood Brewery in Oxfordshire. David did much of the restoration work on the 1951 Bedford OLB.

A·Z OF LORRIES

In model form
By Mike Forbes

This, the fourth part of an occasional series of articles, looks at vehicles which have been modelled over the years. For some readers it will be a trip down memory lane. For others it might be enlightening to see just how many - or how few - models have been made of their favourite lorry.

More variations on a theme, the Corgi Classics Bedford 'O' as an artic, box van and Luton van - usually wrongly called a pantechnicon by most collectors.

'You See Them Everywhere'

We've covered the 'A's - AEC, Albion, Atkinson and Austin - not necessarily in the right order alphabetically, but never mind. Now what do you think B is for? The clue is in the heading - 'You See Them Everywhere....'

Yes, it's the old saying we all remember. And yes, we're still talking about Bedfords - but this time small ones and the saying is still true. It might be over ten years since the last real Bedford hit the road, but literally thousands are still being made in model form all the time - not to mention all the older models there are still about to collect.

The model scene always seems to mirror reality. These days, there are many old Bedfords being preserved and rallied by their proud owners. Many of them still in the transport business, remembering how they drove Bedfords in days gone by. Sometimes the Bedford was the first lorry they drove and they like to look back affectionately on 'the good old days'.

Many more would-be preservationists aren't able to own and run the real thing,

Different 'O' model Bedfords - left to right, the Dinky tipper, the Matchbox furniture van and 'wreck truck' and the Corgi Classics box van.

so the next best choice is to have a model, or maybe a fleet of models. So what's available? When it comes to Bedfords, the answer is lots - although there are some significant gaps, as we shall see.

Where do we start? The obvious way is to start with the smallest and work our way up to the biggest. But do we do this in terms of the size of the model or the prototype? Let's be awkward and opt to work our way up through the different prototypes in terms of their weight ranges.

We start with the Bedford K, the little 30 cwt truck so beloved of operators like builders, coal merchants and endless small businesses who needed a vehicle during the course of their operations - the sort of companies which later used a 35 cwt Transit, but that's another story.

Unfortunately - wallet-wise - we start with a bang. Just before Models of Yesteryear went 'mail-order only' and hiked the price, a very nice little Bedford was produced in about 1/43 scale. It was a typical little market-trader type lorry, but was lettered for George Farrar and came with a bag of genuine Yorkshire stone.

Three lorries based on the same chassis, the Dinky dustcart, tipper and dropside artic.

Trouble was, supply never met demand, right from the start, so it comes with a premium price - up to £30. Ouch. Later issues have been of rather unlikely fire engine versions and so on. A pity.

Still, coming back down to earth, there's always the Lledo version. This comes in many different liveries on several different versions. There's a box-bodied version, plus an integral parcel type van which also comes as an ambulance version with side windows. All quite convincing but to an awkward scale, something a little smaller than 1/50. It's a pity Lledo hasn't yet made a dropside version, although it is possible with a little ingenuity to combine parts from other Lledos to make your own. ●

The rare - and expensive - Yesteryear Bedford K, with examples of the three different types of model of the same type of Bedford from the Lledo Days Gone range, box van, parcels van and ambulance.

Continued next month

CLASSIC COMMERCIALS and Vintage

POSTERS OFFER

Magnificent original commercial vehicle 50s, 60s and 70s colour ads at large A3 size (16"x11") encapsulated in clear plastic for complete permanent protection.

Twenty Two famous commercial vehicle colour ads, to transport you back to the golden age of classic commercials. Printed on high quality art paper encapsulated in a plastic laminate to give them a permanent beauty. Available to readers of *Classic & Vintage Commercials* magazine. **PRICE £6.95 each.** (All prices plus £1 P&P per print). Please allow 21 days delivery.

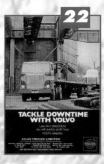

Prices correct at time of going to print

Model Bedfords

The Dinky Toys 'Big Bedford' was the first model S-Type, seen here as the later Heinz van - but not the most desirable tomato sauce bottle version - and the Corgi Major Toys S-Type low loader - note the later type of grille.

After being deluged with readers asking for the return of a regular feature on models, we've dusted down Mike Forbes, editor of Diecast Collector magazine, and overhauled him for further service. This month, Mike looks at new releases, and returns to his study of Bedford models through the ages

You see them everywhere! Once it was true of the real Bedford lorries, but now it's just true of models representing the marque, which are keeping the name alive among enthusiasts.

For those of us who can't preserve a bit of history ourselves in the form of a real lorry to take to rallies on summer weekends, the next best thing is a fleet of models, and if Bedfords are your thing, there's plenty to choose from.

We have previously looked at the lighter end of the Bedford range - all the normal control models from MLD to O-type and later A, D and J-Types.

Now it's the turn of the generally heavier forward control models, starting with the one they called the 'Big Bedford' right from the start.

Mind you, in today's terms it looks positively tiny - how times change!

S is for... the Bedford S-Type. Even my kids know what one of these looks like, as there are so many preserved examples around

(many having had an easy early life as Civil Defence vehicles) as well as lots of models.

The first model S-Type was introduced by Dinky Toys back in the 1950s as the 'Big Bedford lorry.' The dropside usually came in maroon and brown or blue and yellow, but if you're lucky, you might find the rare pink and cream version.

Far more valuable, if it's a mint and boxed example, is the Big Bedford Heinz 57 van in the Dinky Supertoys range, especially if it's the version with the tomato sauce bottle on the side rather than the more common baked beans can.

Next, in order of age of the models, come the various versions of the Bedford S from Matchbox Toys. It started with a tiny (about 1/100 scale) low loader. This was usually a green tractor unit with a fawn trailer, but there were very few two-tone blue versions made before a larger, and, as Lesney saw it, better value version, again in green and cream, at about 1/90 scale.

Also in this scale, described at the time as 'TT', were a Bedford S-Type tipper. This had the characteristic rounded steel tipper body of the time and a compressor lorry. This was

less realistic, as the Big Bedford would have been a bit over spec'd for such a job.

Another TT scale model was a Matchbox 'Major Pack' featuring a Bedford S tractor unit with a Wall's Ice Cream refrigerated trailer. Imagine how imposing this would look in 1/50 scale! But staying with Matchbox, an even bigger S-Type artic was its car transporter. This was mainly produced in blue, but again there were a few red and grey examples towards the end.

One tiny and treasured model in my collection is a Benbros Bedford S artic van, in the same scale as the original Matchbox low loader. Benbros also made a little artic Esso tanker, and it's a lovely little thing, if you can ever find one!

There were a couple of S-Types from Budgie Toys, with an unlikely bulk tipping body on a six-wheeled chassis, labelled Ham River Grit, and an even less likely enormous dumper body, which in real life would probably have buckled the chassis before the truck was even loaded!

Staying with older models, Corgi Toys made a Bedford S-Type tractor to pull its Carrimore car transporter, and a choice of two types of low loader. This was a model of the later S-Type cab with the grille above the radiator slats, rather than the plain panel with the chrome Bedford badge.

Right to left, the Benbros S-Type artic and the Matchbox line-up - the first low loader, the second bigger low loader, the tipper, the Walls ice cream artic and the car transporter.

Up to date with Corgi Classics, the S-Type artic of Jack Richards, the dropside of Ken Thomas and the Spratts van in the Golden Oldies range.

Coming up to date, the Corgi Classics range has featured the Bedford S-Type in a number of guises. A few years back, Corgi released its Golden Oldies range, with vans in the old liveries used by Dinky back in the 1950s. The Bedford S took on the colour of Spratts, Lyons Swiss Rolls and Weetabix - all Guys back in Dinky's day.

Then there's the dropside four-wheelers from Corgi, in Ken Thomas, W&J Riding and Pickfords colours among others, plus those with a tilt added in AFS and Teltey's Brewery livery. Artic flats in BRS and Jack Richards colours complete the scene, but the less said about the Shell-BP artic with the tanker trailer off the Scammell Highwayman the better. The four-wheeled tanker in Esso livery is much better.

Last of the Bedford S-Types to appear came from Lledo in its Vanguard range. The lorries are made in the (at least outside America) usual 1/64 scale, but the models don't fit in with any other makes.

The Vanguards Bedfords have appeared as box vans in a variety of colours: Heinz, of course, Kodak, Post Office Supplies and others, including a set of two Ken Thomas vehicles. We have also had tankers, with rather oversized tanks, in Mobilgas, Shell-BP, Texaco and other colours; and most recently as a dropsider in a set of two lorries in Whitbread livery.

The Bedford S in different scales. Left to right. 1/50 Corgi Ken Thomas dropside, 1/76 Marquis Models white metal cab conversion on an EFE chassis, 1/90 Matchbox tipper and tiny 1/150 Matchbox low loader.

Lledo's Vanguards Bedford S-Types - the Shell-BP tanker again, and the Heinz van.

Appliance of Excellence

Fire engine restorer Mick Paull has meticulously restored one vehicle after another, and the latest project, a Bedford MSZ, is probably his greatest achievement so far. We meet him.

One of the star restorations to appear last year was Mick Paull's incredible effort on this 1940 Bedford MLZ fire engine. (Photo: Simon Rowley)

'**Err, having restored a series of fire engines to what in many people's eyes is perfection, does it get any easier with every one you do?**"

"No," comes the simple answer to our question from Mick Paull. "I learn a lot with every restoration I do."

Looking at some of Mick's completed projects you'd certainly think he'd arrived at the end of the learning curve long ago.

Mick, one of the main organisers of the much-renowned Odiham Fire Show in Hampshire, and stole the most recent event with his own latest restoration, a 1940 Bedford MSZ, transformed from a chassis, scuttle and pile of rot.

APR 679 was built as a fire engine by Superline Bodies, based in south east London. It was originally fitted with a 1000gpm Pulsometer pump, replaced by a Dennis No 2 pump in the 1950s.

After service with the National Fire Service, the MSZ passed to the Dorset Fire Brigade in 1948, and retained as a front-line machine until the early 1960s. It was then loaned to

A nice surprise for the neighbours. The Bedford (or remains of same) as purchased. (Photo: Mick Paull)

Now running for the first time in 20 years, and waiting for the body to be rebuilt. (Photo: Mick Paull)

Bedford is equipped with Pulsometer pump, bought from a clearance sale and almost exactly matching the unit which would have originally been fitted. (Photo: Simon Rowley)

the National Trust to provide fire cover on Brownsea Island, the birthplace of the scout movement, in Poole Harbour.

After about three years, it was sold off and fell into the hands of a group of 1960s students who took it to carnivals and used it for other merrymaking. Needless to say, the maintenance schedule wasn't meticulously adhered to, and the by then much-abused Bedford was saved from destruction by a Burlesdon, Southampton-based Austin Seven enthusiast, then to well-known fire engine preservationist Colin Mockford.

Eventually the Bedford, or what remained of it, was moved to Mick's workshop in Hampshire,

"It was basically a chassis and scuttle, there was nothing left of the body," recalled Mick. "Just about everything that could have been seized was seized, apart from the axle.

"Being on Brownsea Island, the salt air did the body no good, and the scuttle is steel. I've seen a short film of the Bedford when it was owned by students, and I think they must have been having a competition to get as many students on it as possible, like there was a fad for squeezing as many people as possible into a Mini at the time."

The Bedford's engine had been left idle for 20 years and was totally rebuilt, with the crank being reground. The brakes were also completely overhauled.

The bonnet, wings and front scuttle were original but needed much repair. The gearbox was rebuilt and, with the electrics being non existent, it goes without saying that a complete rewire was the only option.

A complete new ash-framed body was rebuilt, using some of the original timber. The vehicle was then resprayed.

A replacement Paulseometer pump was bought from a dispersal sale in Cambridgeshire, and closely resembles the original, and the ladder actually came from Odiham Fire Station, having served on several generations of vehicles and only recently retired

The result of this two year restoration is a stunning fire engine, which as Mick says: "nips along nicely and will stop on a sixpence."

Mick's major previous project was a 1939 Leyland FK9 Cub, FLJ 356, supplied to the Bournemouth Fire Brigade and passing through the hands of several preservationists before Mick bought it in 1995.

Looking at the 'before' photos of this Leyland, you'd think the Leyland was in much better condition when it arrived, but Mick puts us right. "It was similar to the Bedford, but more complete," he recalls.

The limousine body was almost a kit of parts, which required piecing together like a jigsaw, with sections of the rotten roof renewed.

The Leyland 29.5hp straight six petrol engine just would not tick over properly, a situation which had baffled previous owners. The condenser was an obvious culprit but the problem turned out to be down to faulty low tension leads. There were also serious problems with the gearbox, which had to be rebuilt, the first motion shaft bearing having broken up. The clutch was professionally rebuilt.

At the time of purchase, the Leyland was fitted with a Merryweather war-pattern 50ft escape ladder with metal wheels, and a similar unit was bought with

Sea air had helped reduce the Bedford's bulkhead to this state! (Photo: Mick Paull)

one good example being made from the pair. Yet more restoration will see the Leyland equipped with a Merryweather lightweight trussed pattern escape very similar to the one it carried when new.

Thankfully, the Gwynne 700 gpm pump which came with the Leyland was in good condition thanks to work carried out by the previous owners, and just needed some adjustments.

Having said that, the whole project took 1600 man hours over seven months, but says Mick: "It was well worth the effort."

One of the first vehicles to join Mick's collection was a 1942 Austin K2, GLE 973, which started off as a National Fire Service heavy unit but was later converted into a major pump with a semi-limousine body, which gave a small amount of attention to the fire crew at the back. This work was carried out by the Berkshire and Reading Fire Brigade Workshops in the early 1950s. A Dennis No. 2 pump was fitted immediately behind the front crew department.

Stationed at Hungerford, Berkshire, for many years, the appliance was named John O'Gaunt II after the powerful 14th-century nobleman. It was sold out of service to Adwest Engineering on the Woodley Aerodrome near Reading, coming to Mick in a 'rough but complete' state in 1978, then receiving the 'Paull perfection' treatment.

Another 'grey machine' which has been totally restored by Mick is a Fordson 7V heavy unit, GGJ 12, supplied to the NFS in 1941.

Its service history is not known, but the Fordson originally appeared on the preservation scene in the early 1970s, and was in the care of several owners before it came to Mick in a state of disrepair in 1992.

Within two years, and following a total rebuild, it had been restored to such authentic displays that it was able to take its rightful place alongside other NFS

Bulkhead rebuilt, chassis blasted and repainted. (Photo: Mick Paull)

Paulseometer pump bought for the Bedford was stripped for restoration. The cast iron casing was found to have a large split in it, probably caused by frost, which had to be welded. (Photo: Mick Paull)

appliances in displays of World War 2 firefighting, presented by members of the Fire Service Preservation Group and a very popular arena event at rallies.

Despite the extraordinary detail and standard of his restorations, Mick isn't a professional restorer but a lorry driver by trade. As he says, he's learnt as he's gone along.

Any more restorations planned? "Eventually, but for the moment I'm going small scale and building a model Scammell," he said.

So, can Mick sum up the appeal of elderly fire engines? "I think it's the fact that you have a vehicle which can actually still perform its original duty, and could still be called on to do so.

"And, of course, they bring out the kid in you!"

Mick Paull with another of his immaculate restorations, this 1939 Leyland FFK9 Cub. (Photo: Simon Rowley)

Leyland roof needed major patching. (Photo: Mick Paull)

Leyland woodwork underway. A lot of rot had to be dealt with. (Photo: Mick Paull)

Leyland as acquired - in a much worse state than it looks here. (Photo: Mick Paull)

Mick's 1941 Fordson 7V heavy unit seen at an Odiham Fire Show. The Thunderbird 2-like item in the background is a mock-up of a wartime V2 rocket, by the way. (Photo: Simon Rowley)

Bedford MLZ reunited with some of the many people who remembered it throughout a long career. On the left is Tom Porter, who with his mother, Fanny, bought the vehicle new in 1939.

TIME AND PLAICE

Eddie Doig's restoration of this wonderful 1939 Bedford MLZ fish and chip van resulted in a 750-mile trip, including nostalgic returns to its former Norfolk home and the factory where the body was built... plus reunions with 40 people who remembered the vehicle in service!

"**W**e lit one of the fires, stood back and just let the 40 people present reminisce as the smoke wafted around us." What an ultimate reward for the painstaking restoration of a 1939 Bedford fish and chip van, taken back to the area of Norfolk where it had served the hungry for so many years.

Restorer Eddie Doig had promised he would make the trip, but he had no idea so many people would want to see the M-type again. "It was amazing. Apparently there had been an article in the local paper, and so many people had fond memories of the van," Eddie says.

Most hadn't see the vehicle since February 1971 when owner George (Paddy) Staines, who lived near Kings Lynn in Norfolk, decided that no way was he going to get himself involved with new-fangled decimal currency. "He drove the van into a lean-to and that was it," said Eddie.

Bedford restored very much to as-in-use condition. Two of the chimneys are condensers.

The Bedford remained there for almost 20 years before being sold at auction for the grand sum of £100 in 1990. The lady buyer kept the vehicle a few years, but did nothing to it.

Eddie, from Wirral, Merseyside, bought the vehicle from a dealer in Huddersfield, West Yorkshire.

"I'd actually been going to restore a Bedford OB coach. Then I received a call from the dealer with the M-type. He'd described it as an ex-British Railways vehicle as it had been fitted with doors from an ex-BR Scammell Scarab, which had numbers on them," Eddie remembered.

The vehicle was certainly in a battered state. "It was in poor condition and had been vandalised with the instruments smashed. But it was otherwise intact. That's why I bought it. It was like a time capsule. There was still fat in the fryers."

And so the restoration began. Eddie decided to send the 3519cc petrol engine away for overhaul, and this was carried out by Birkenhead engine centre. The wooden framework was well past its best. "Not only was the wood rotten, but I've never seen so many woodworm in my life," said Eddie.

Ash was used for the reframing, except across the back of the Bedford where there's a piece of teak. Repanelling was also necessary. The cab floor was replaced, and the rest of the floor treated as this still carried its original lino.

A complete rewire, brake overhaul and new kingpins were all needed before the van eventually passed its MoT. The roof, originally canvas, was replaced with vinyl.

All four chimneys were also renewed – two are actually condensers while the other pair get rid of the smoke. The original coal-fired range, built by a firm still in

The Bedford MLZ when almost new, though note how scratched the sides are, presumably from hedges along the narrow country lanes where she worked. You can also see the white on the left, useful in the wartime blackout. Ready to serve you are, Fanny Porter and Miss Haskett, who later married the Bedford's last operator, George Staines.

business, Preston and Thomas of Cardiff, was retained. Firelighters were used to get rid of that vintage chip fat!

Meanwhile, Eddie decided to do some simple detective work. "When I got the Bedford, there was no paperwork. However, in the cab was a small aluminium plate on which, painted by hand, was the inscription: 'GW Staines, K Lynn, 30/weight, 2 ton 17cwt'."

Eddie checked the King's Lynn phonebook in his local library and found a listing for Staines. "One call and I hit the jackpot. I was put in touch with Mr Staines' daughter, who used to work on the van, and then 'Paddy' Staines himself. He and his family were a great help with photographs, information and enthusiastic support. I promised them I would take the van home when it was done."

'Paddy' proved to be a mine of information. Through him, Eddie was able to track down Tom Porter, who with his mother, Fanny, bought the van new in 1939. It then passed to Jim Haskett, who

750-mile round trip also saw Bedford introduced to rallygoers.
Here it is at the Great Dorset Steam Fair, having received the award for best commercial, and below, on another occasion, at the Malpas, Cheshire, show.

Bedford outside the very chippy where its wares were prepared in the 1960s. You can just smell that aroma!

The Bedford as acquired, suffering from vandalism, weathering and woodworm – definitely a 'battered' state!

Back at the former AW Watkin workshopin Bigg

Engine is out for rebuilding, and wings and bonnet removed ready for the welding torch.

An enormous amount of woodwork, and woodworm culling, was needed on Bedford.

drove for Mrs Porter during the war. Mr Staines bought it in 1963.

Throughout its career, the van had served villages to the south of King's Lynn, including, during the war, a prisoner-of-war camp. The prisoners used to swap goods they made for fish and chips. In fact, during the restoration, Eddie found a home-made cigarette lighter down one of the window pillars, which almost certainly came from the camp.

But the Bedford was not without a few near misses. One Friday in 1967, it completed its round despite deep snow. Sadly, only a mile from home, a simple misjudgement of the road led to the truck teetering over the edge of a ditch. In that precarious position, it could have all too easily slid into the ditch at any time. Broken lemonade bottles lay all over the floor, but of greater concern were the two coal fires and several gallons of boiling beef dripping sloshing around in the pans.

The situation looked bleak but then, out of the darkness, a breakdown truck miraculously appeared. In a short time the Bedford was back on the road, the driver refusing all payment for his services.

Back to today and Eddie decided to take the Bedford on a 750-mile, 10-day tour of nostalgic reunions and vehicle rallies.

"The first day began bright enough, but we were just past Chester when it began to rain. Half an hour later, the first drop of water landed on the back of my wife, Iris' neck. Half an hour after that, we were both sitting in our waterproof jackets with the prospect of another four hours' driving ahead of us."

Undaunted, the pair attended the first port of call, last year's excellent Bedford Rally at Somersham, Cambs. "One of the penalties of owning a fish and chip saloon is that nearly every conversation beings with someone saying, 'Six of chips, please.' Somersham was so different in that we were among like-minded enthusiasts, and we talked to some really nice and informed people during the day."

...eswade, Beds, where the body was built.

Restoration is well underway, with framing and repanelling complete.

Then it was over to Norfolk for the big reunion, the van being photographed in its old haunts and many stories shared until darkness fell. Unfortunately, 'Paddy' died two years earlier, but there were several members of the family present when the Bedford arrived at his daughter's home in St Germans.

After this it was down to Biggleswade, Bedfordshire, to the only recently-closed showroom of AW Watkin, who built the chip van body. The Bedford was delivered to them as a chassis and scuttle.

Eddie drove the MLZ around to the soon-to-be-demolished workshop from which it had emerged 61 years earlier –

Bedford still contains original coal-fired range, built by Preston and Thomas of Cardiff.

and never been back since.

Finally, it was a long trip south to the Great Dorset Steam Fair where Eddie and Iris were to spend the next few days.

"Everything was fine until I went to get one of the camp beds out of the back of the van, only to discover it had been resting against the hot chimney at King's Lynn and had a big black hole burnt in it. Fortunately it hadn't burst into flames as our triumphal return to Norfolk could have ended there and then with the Bedford ending up as a pile of ash," said Eddie.

The Bedford returned home having completed its long journey. "In all those miles, she didn't miss a beat. She didn't miss many petrol stations either, but that unfortunately goes with the territory," said Eddie.

There were many enjoyable aspects of the trip: sharing reminiscences and giving a lot of people great pleasure to see the Bedford – both rallygoers introduced to it for the first time and those who remembered it in use all those years earlier. But there was another excellent surprise for Eddie and Iris.

"To my utter amazement and delight, the Bedford had been judged the Best Vehicle in show at the Great Dorset steam fair. What particularly thrilled me was that she had won not because she was the shiniest or prettiest vehicle on show, but she was judged on originality.

"I presented her as the working vehicle she was, faults and all, and it's a poke in the eye for the chromed wheel nut brigade that an ordinary van can take such a prestigious award."

SUBSCRIBE TO

S Class

Bedford S-types helped build Knowles Transport into one of today's most prestigious hauliers. Now the company has restored two examples, one bought new. Nick Larkin reports.

Photos: Martyn Barnwell

Winning one of the haulage industry's most prestigious awards, Motor Transport's Haulier of the Year 2000, is no small achievement. Knowles Transport's immaculate fleet of 90 Volvos and Scanias is seen over a wide area. Its two vast Cambridgeshire sites incorporate more than a million square feet of warehousing, and there's even a food-packaging department.

Yet this long established company has much to thank a relatively humble lorry, which played a major part in its development over the years – the Bedford S-type.

More than 40 were operated in the 1950s and 60s, a time when the firm was expanding rapidly. Today, two examples, a 1959 Scammell trailered example bought new and a 1954 four-wheeler new to another Cambridgeshire haulier, have been restored from hideous wrecks to immaculate condition.

Managing director Tony Knowles said:

"The S-type played an important part in our development. With our two restored examples, I was trying to recreate the vehicles my father had in the 1950s and 60s. We must have run 40 examples."

Tony's father, Gerald, started the firm when he was 17, in 1932. He ran one vehicle, a two-ton Ford, putting sugar beet on to the railway stations. The 1940s was a period of expansion. During the war, he used to take caterpillar tracks all over the country working for a contractor in Manea, Cambs, Derek Crouch, who used to build aircraft runways.

After the war, Knowles' work increased as the company transported bricks from Whittlesey and Peterborough as well as taking vegetables to market.

Recalls Tony, who now runs the company with his brother, also called Gerald: "When we got nationalised, BRS took the vehicles away. Father then had four new Bedford O-types, three four-wheelers and one Scammell-trailered artic. He started doing contract work for the Whittlesey Central Brick Company."

Department S! 1959 artic and 1954 four-wheeler, restored from wrecks. What a difference some chrome makes, otherwise little difference to S-type styling over the years.

You'd never have thought the S-type was a cut-down farm hack until rescued for restoration.

Tel. DODDINGTON 233

S-Type dash almost echoes that of contemporary car.

Diesel

Above and below, a local legend! Deliveries Dial Doddington earned the affectionate 'Big D' nickname.

DELIVERIES
DIAL 233/4/5
DODDINGTON

BEDFORD

MEW 841

KNOWLES TRANSPORT

Tel. DODDINGTON 233

KNOWLES TRANSPORT Ltd
Haulage Contractors
WIMBLINGTON · CAMBS

BEDFORD

KEB 519

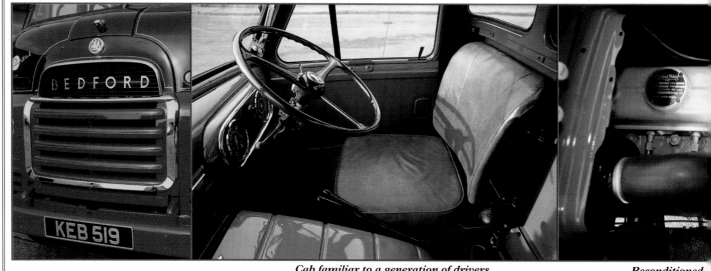

Cab familiar to a generation of drivers.

Reconditioned

When the haulage industry was denationalised, Knowles tendered for several Cambridgeshire BRS depots and got Ramsey, March and Whittlesey, including the vehicles and A-licences. "There was some old stuff including Maudslays, but my father transferred the A-licences to new Bedfords. The marque was always his favourite," explained Tony.

"As soon as I could walk, I was in the garage and under lorries, helping to grease them. They were part of my life. Bedfords were all father could afford in those days.

"They were good for the job – a seven-ton Bedford always carried ten tons. Father always ordered them on 900x20 tyres, with extra leaves in the rear springs, and off they went as a ten tonner.

Regulations were different then!"

The Suez crisis meant Knowles senior was keen to get hold of diesel vehicles. "He tended to keep the Bedfords for four years, and was very keen to sell the petrol examples brought from 1952."

Then two S-types with Perkins R6 engines arrived. "It wasn't a very successful engine, but it was the first diesel you could get in an S-type. In fact, it was all you could get for some time.

"We went on to the bigger artics, and the Leyland engine was more powerful than the 300 Bedford, which wasn't up to carrying 14 tons on a 12-tonner."

Eventually the S-type was withdrawn from the fleet, but its memory lived on. Then four years ago

Bedfords mean nostalgia for Tony Knowles.

S-type shows off its Scammell trailer.

Founder Gerald Knowles takes centre stage with two earlier S-Types.

S-type artics among this line-up.

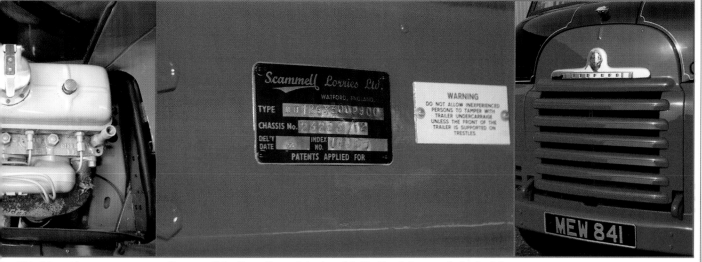

ne fitted in artic

Scammell-built trailer original to this Bedford, but needed rebuild.

New dropside body made for four-wheeler.

Artic's cab was bought from John Mould. Knowles still had some panels in stock.

Engine in place, the four-wheeler waits for cab to be fitted.

Four-wheeler's Perkins R6 engine was built up from two units. Spares are difficult to find.

y Knowles with his brother, Gerald, and a new Vauxhall Victor in 1958. The vehicle just visible over the car bonnet is the now restored artic.

1950s yard shot just after the buying out of local British Road Services depots, including Fodens along with the familiar S-type. Some of the trucks here are ex-BRS.

Artic had original engine in place, despite use as farm trailer.   *Chassis on both vehicles were treated to a good coat of red oxide.*

Scammell trailer as originally found, complete but needing a major rebuild.

came an amazing coincidence. "An owner-driver who was working for us lived at Stoke Ferry in Norfolk, where he found an old Scammell trailer. He suggested we had a look at it. It turned out to be an artic," recalled Tony.

"The tractor unit had been cut in half and used as a trailer dolly. The trailer was there, and the back half of the unit was still underneath it. They'd taken the front off, bent the two chassis rails together, put a pin on and towed it around with a tractor.

"The cab had gone, but the number plate was still on the back, KEB 519. It was one of ours. We had to have it!"

The artic was new to subsidiary company Wisbech Roadways, which still exists.

"Father started it because he was doing a lot of work at the docks in Wisbech and King's Lynn," explained Tony.

The S-type, or what remained of it, was rescued. The trailer was completely stripped, the chassis shotblasted and new side rails were made. The wood on the flatbed was replaced with new. A cab in excellent condition came from well-known classic commercial collector John Mould, a chassis was found in Lancashire, and a reconditioned Leyland 350 engine came from C&G Coachworks in Much Wenlock, Shropshire.

"We had the original, but decided to take the chance of a reconditioned unit," said Tony.

Some panels and other parts for the vehicles, even a steering wheel, came from Knowles' own stores. "We never throw anything away," explained Tony. "We used everything we could from the original vehicle."

The four-wheeler was found in a scrapyard near Earith, Cambridgeshire, and like the artic, its registration number proved to be its saviour.

"The registration was MEW 841, and we ran an S-type registered MEW 636," said Tony. The vehicle is believed to have been new to Prime Godfrey, based at Swavesey, near Cambridge. "It was a decaying wreck, but it was so like the vehicles we ran."

The Bedford needed a complete rebuild, the chassis being, as Tony said, absolutely shot. An ex-Ministry of Supply S-type was secured as a donor vehicle. The Perkins R6 engine was constructed from two units.

"It's very difficult to get parts for the R6," said Tony, "though we did get some pistons and liners."

All the metalwork to reconstruct the body came from Knowles' workshops. Work on this vehicle was carried out by Norfolk-based restorer Graham Whitby, who also advised on the artic restoration. Knowles' fleet engineer, David Butwright, was also much involved.

Both trucks were painted by Norman

Smith of Throckenholt, who paints all Knowles' modern fleet. Two-pack paint was chosen. Signwriting was contracted to Tony Warren, based at Market Deeping, Lincolnshire, whose work also appears on Knowles modern vehicles.

The company has a collection of around a dozen preserved trucks, including various Fodens and a Scammell R8, but seeing the Bedfords finished has been particularly important to Tony.

"The four-wheeler was completed first. It brought back so many memories. The finished article was so authentic, I was over the moon, especially as it was one of our own original vehicles. The trucks are just how they were turned out from the factory in the 1950s, when we got them from Murkett Bros at Huntingdon."

Records show that £1525 was paid for the artic, which was sold in 1956. NEW 812, a four-wheeler similar to the restored example, was bought for £1484 in 1954 and left the fleet in 1960.

Knowles' recent award confirms it's very much an A-class haulier, but the four-wheeler S-type has been restored to recall the days when the company was known locally as The Big D.

The reason? That letter appeared boldly on the back of vehicles, along with the message, "Deliveries Dial Doddington 233/4/5." Since then, Knowles has expanded even more than its phone number!

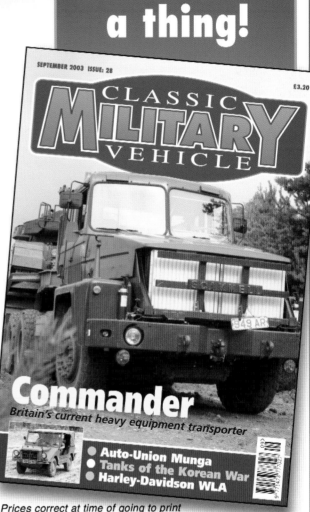

BEDFORD BIRTHDAY!

This year's Bedford Gathering attracted 225 vehicles – including many other makes – to celebrate the marque's 70th birthday. Nick Larkin reports.

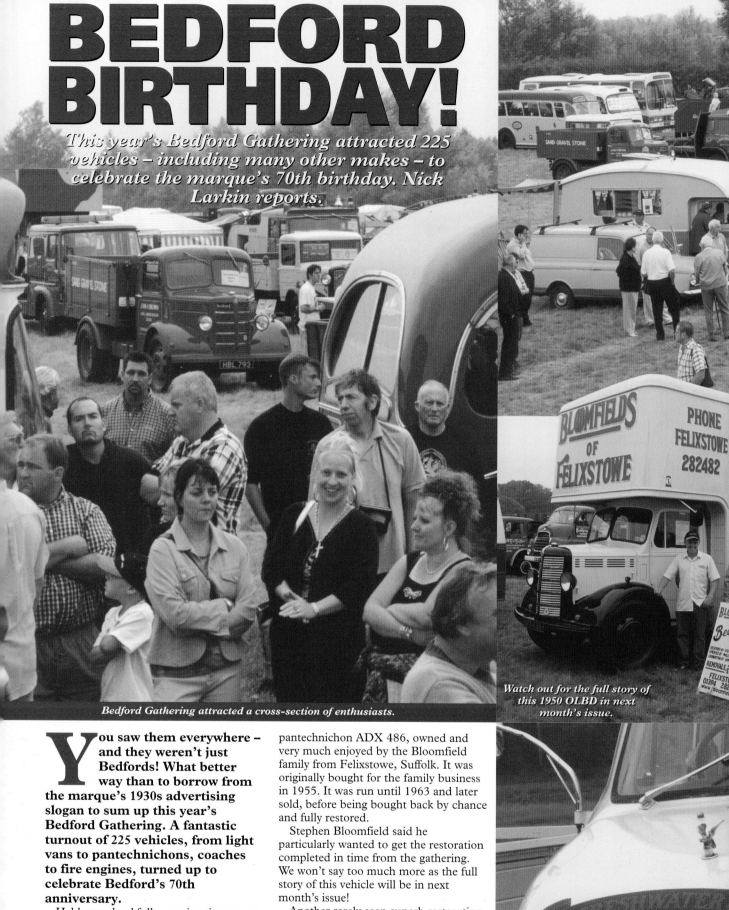

Bedford Gathering attracted a cross-section of enthusiasts.

Watch out for the full story of this 1950 OLBD in next month's issue.

Err, the owners love it really, honest!

You saw them everywhere – and they weren't just Bedfords! What better way than to borrow from the marque's 1930s advertising slogan to sum up this year's Bedford Gathering. A fantastic turnout of 225 vehicles, from light vans to pantechnichons, coaches to fire engines, turned up to celebrate Bedford's 70th anniversary.

Held on a thankfully massive site next to the Post House Hotel in Cambridge, the gathering attracted vehicles from many countries, including a sizeable Dutch contingent.

Classic and Vintage Commercials' publishers, Kelsey Publishing, was among the sponsors.

Attracting much attention was surely one of the year's best newcomers to the rally scene, 1950 Bedford OLBD

pantechnichon ADX 486, owned and very much enjoyed by the Bloomfield family from Felixstowe, Suffolk. It was originally bought for the family business in 1955. It was run until 1963 and later sold, before being bought back by chance and fully restored.

Stephen Bloomfield said he particularly wanted to get the restoration completed in time from the gathering. We won't say too much more as the full story of this vehicle will be in next month's issue!

Another rarely seen superb restoration was 1944 Bedford OWL, BHJ 800, owned by retired garage proprietor Peter Skinner from Raunds, Northants. New to a civilian operator, the Bedford has been finished in the livery of Whitworths, which operated 14 similar vehicles. "It's been restored for six years, but it hasn't been to many shows," said Peter, who also owns a Bedford OB coach.

Restoration of this Bedford took

Smaller Bedfords and other makes played a major part at the event.

TL said to be the last Bedford to leave Marshall's factory is now for sale.

Multi-coloured, cosseted J-types.

around four months of intensive work, with Clive Wood of Rushden carrying out the superb signwriting.

A surprising appearance was made by a TL chassis and cab which could be the last Bedford of all. "We're still not sure if it was the last built or the last to leave Marshall's Cambridge factory," explained owner Tony Croft of Uxbridge-based Bedford specialists Croft Brothers (UK) Ltd.

"We were pleased to secure this vehicle. One idea was to convert it to a recovery truck, but it's only Euro 1 specification and now for sale. It'll probably be exported."

The 1997 vehicle had travelled a mere 280km when it was displayed at the rally. "Half of those were coming up here," said Tony. Phone 01895 850700 if you're interested in the TL.

Both the Best in Show and Best Bedford Commercial in Show award winners were the same as last year, these being Geoff Heal's pristine 1951 Duple-bodied Bedford SB coach, PPH 698, and GSJ 761, the 1951 Bedford OLBC dropside of Oxfordshire-based Wychwood Brewery.

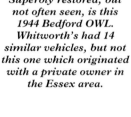

Superbly restored, but not often seen, is this 1944 Bedford OWL. Whitworth's had 14 similar vehicles, but not this one which originated with a private owner in the Essex area.

The gathering started in 1998 with only 30 vehicles. After last year's rally, it was decided that the event had outgrown its original site, Dews Coaches' Somersham premises.

Commenting on this year's show, which was raising funds for the Mid-Anglia General Practitioners Accident Service (MAGPAS), Dick Haughey of the organising committee said: "It's been wonderful. We must have had about 225 vehicles, and we're hoping to hold an even bigger event here next year."

Still in regular use is this 1965 Bedford TK.

1944 OYD flatbed was just one of the line-up from enthusiastic Dutch operator Jan Geffen of Hedel, Holland.

TK took long-distance award at Bedford Gathering.

STILL REMOVING

A recently established family-owned removals business is getting excellent service from its much-cherished 1982 Bedford TK pantechnichon.

The Marsden-bodied vehicle is in use six days a week by Okehampton Removals and Storage, based in Devon. CBA 357Y is pictured at the Bedford Gathering, where it took the award for the vehicle travelling the longest distance to the event.

The TK has a Bedford 330TD engine and a five-speed gearbox. It was previously operated with Ray Kirkham of Oswestry, Shropshire, and before that, Campbells of Croydon.

Mike Griffin started Okehampton Removals and Storage last November, primarily as a family business for his son Christopher, currently 11, to go into.

Mike had previously been involved in the removals business for some years after leaving the RAF, working for the now defunct Euromove, though more recently he was in charge of security at Noel Edmonds' country estate. He now employs two people in the new venture, the fleet also including a Bedford CF Luton and an LDV 400.

"I've always been a fan of the Bedford Marsden and this is a lovely truck," he said.

"We haven't needed to do any major restoration on it, just spray it in our colours. It's in use six days a week doing long-distance removals. It's just been from Devon to Sunderland. I'd particularly wanted to take it to the Bedford Gathering, despite the five hours or so journey."

Weighing 12.5 tons gross, the vehicle has a twin-sleeper cab.

Adds Mike: "We generally go about 55-58mph. Because the new trucks have speed limiters, it's not so bad. It tends to be a bit slow on hills, but on the flat, it's a lovely cruising machine."

Bedford works five days a week, sometimes six.

Family owners, Mike Griffin, wife Jan and son Christopher, 11.

Little work was needed apart from a repaint in Okehampton's attractive colours.

REMOVING ST

A 1950 Bedford OLBC pantechnichon has been restored by the family who bought it as a five-year-old and have a rich history in the removal trade. Nick Larkin meets three generations of the Bloomfields, the Bedford, its former driver and a unique Ford restoration project.

(Photos: Glyn Barney)

There's no time to dilly dally on the way if you want to follow one removal van – or rather its colourful history! The fascinating journey involves four generations of the Bloomfield family, and as far as the vehicle is concerned, long service life, a lucky survival for a most unusual of reasons, and then eventual sale. But, unlike the distressed narrator of ye olde music hall song we borrowed from earlier, this 1950 Bedford OLBC, registration number ADX 486, has found its way home, and is now kept in the same shed where it spent the latter half of the 1950s.

Where to begin? Well, 1916 would be a good time. John Bloomfield, commonly known as 'Jack', was invalided out of the First World War. Unable to find a job, and not one to mope around, Jack decided to go into the removal business, beginning with a handcart. Bloomfields of Felixstowe has been in business ever since.

A couple of years later, Jack bought his first lorry, which is believed to have been a wartime Overland, and around 1920 the business was expanded with a furniture shop, still going strong today.

Later came a 1929 Chevrolet LQ six-wheeler, followed by a string of Bedfords, which served the firm well.

Jack's son, Stanley, now 78, was, as he says, "born into the business" – and remains in it!

"I packed my first box of China when I was nine," recalls Stanley, "and I still have a small involvement. Sometimes they call me out to repair a bit of furniture, another part of the business, or anything like that."

Back to the vehicles. "We had one

Bedford W, which we bought off a firm called Osborne, that being PV 3699. It was a flat radiator grille model, which to all intents and purposes was a Chevrolet. That was in 1937. I drove that myself for 250,000 miles. After the Second World War, we fitted an O-type front on it to make it look more modern."

That vehicle was number four in the fleet, number five being an OLB, FRT 250, bought new. "We subsequently broke that one up and the doors are still around somewhere."

'Jack' Bloomfield died in 1952, and Stanley took over the business, his brother, George also being involved.

In 1955 came the chance of a bargain, which would become number six in the fleet. ADX 486, a Bedford OLBD had been new to the Ipswich firm of Lidbetter's. "I think Lidbetters was giving up at the time, and this vehicle was only four or five years old. It was worth

Restored and signwritten, the finished 1950 Bedford OLBC screams nostalgia!

Bedford alongside the workshop where it was garaged during its service days, was restored and is now kept.

Cab refurbished and seats upholstered by Steve Bloomfield's partner, Michelle.

buying," remembers Stanley.

There was one major disadvantage. "The Luton body was raked forward at the front. That was a useless thing as it reduced the load space considerably, so I modified it myself in our workshops."

ADX settled down to hard, reliable service, travelling some 50,000 miles with Bloomfields. "It did about 15mpg, which is very good *(incredible, considering it's petrol – Ed)* and we had no trouble with it."

In 1966, the OLBC was replaced by a Ford D-series, HRT 477D, which today awaits restoration, and was probably unique even when new. "I insisted on petrol, and they had to get a straight six engine from Canada. The suppliers got the engine. The innovation on the engine was that it had automatic oil valve lifts, which was nothing but a nuisance all the time we had it. It used to start up like a box of nails until the oil had got through," Stanley recalls.

Today the fleet consists of a Bedford CF

Open and shut case. Bedford shows off its 1000cu ft loading capacity.

Luton, bought new in 1985, C529 UDX being about to be restored for further use, and a Mercedes 1317, G235 VEE, which is shortly to be replaced.

But what happened to the OLBC? Stanley explains: "It was put aside. My policy was never to get rid of anything. I always kept them and broke them up. My argument was that if I sold it, it would work against me."

Okay, but why wasn't the Bedford broken? Well, you couldn't have made up a more unusual reason! Stanley again: "We put a large sign at the front of the premises, and the council said we had to take it down. So instead, we put the Bedford there, partly to annoy the local councillor, we must admit."

There it remained until 1986, when Stanley sold it to removals firm Greens of Stowmarket. But we'd better catch up with the family history before going much further.

By then, Stanley's son Paul was in charge of the business, much of his work today involving the furniture shop. But it was grandson Steve, now at the helm of the removals side, who missed the Bedford most.

Now 30, Steve says: "When I was a kid, I used to play in it. My grandad sold it. I didn't like it being sold. I always thought I' get it back if I was patient enough."

And he did, after a chance meeting with the Bedford's then owner, Bill Green, who had restored and rallied the vehicle.

"I was doing a removal up the road and Bill popped in. He said he wasn't using the Bedford and if I made him an offer, I could have it back."

An offer was accepted. Remembers Steve "I bought it back one weekend, and by the next, with help, I had the body off. We did this by jacking the whole thing up, undoing all the bolts and set the body onto oil barrels, lowering the chassis and driving it out."

Well, when we say 'drive,' the engine was hardly in the best of conditions. "The big ends had gone and the

OLBC with the modern fleet of Mercedes and soon-to-be restored Bedford CF.

BLOOMFIELDS OF FELIXSTOWE

BLOOMFIELDS OF FELIXSTOWE

BLOOMFI of FELIXSTO

on the move s

BEDFORD

C529 UDX

ADX 486

G235 VE

MEMBER OF SPEED SKILL NAFWR SAFETY SERVICE

Wonde peri trans foun cupbo

Sister Bedford APV 534 after years in the open. ADX 486 is in the background prior to being sold in August 1986.

Work under on cab and b

Four-litre Bedford petrol engine substantially rebuilt.

Bedford returns to its shed. Note the weather vane!

The Bedford, and from left, Stanley Bloomfield, who bought it in 1955, his grandson Steve, who now runs the removals business and masterminded the Bedford's restoration, and Reg Baldwin, its driver for many years.

pistons were slapping about," recalls Steve. "My granddad had bought a brand new cylinder head for it many years ago, and this still in the workshop."

The crankshaft was reground, and the engine resleeved, rebored and treated to new pistons. A new carburettor was located. The gearbox also needed considerable treatment, with new selector forks.

The chassis was found to be in good condition, and the cab wasn't bad either, having to a degree been protected by the roof over the years. The same can't be said for the doors, which needed a lot of work. New bottoms were gas welded on by a friend's father.

The 1000cuft body and the bonnet needed much work, dent removal and filling being on the agenda. A new grille was found via a specialist in Ipswich. "The original had rust beyond filler – it was terrible!"

The electrics had survived incredibly well, but were thoroughly checked and replaced where necessary. The brakes were stripped and carefully reassembled. The wheel hubs were replaced, as were the window rubbers. Steve's partner, Michelle, reupholstered the seats.

The vehicle had been brush painted, which meant a lot of work with Nitromors and scraper. The respray and much of the bodywork were entrusted to Malcolm Lockhart, who had done coachpainting and body repairs for many years. He now works for James Neill of Stowmarket.

After much work and anguish, an immaculate Bedford emerged, and it was time for those finishing touches. A tremendous job of signwriting was completed by Sam's Signs of Ipswich, its effect being enhanced yet further by some period National Association of Furniture Warehousemen and Removers transfers. "I found those in a cupboard," says Steve.

He says: "At times the restoration was daunting, and I'd sit there thinking how I'd got this to do and that to do, but the enthusiasm keeps you going. I seemed to fall

"You're only supposed to blow paint on the bl**dy doors," as Michael Caine might have said.

Unique petrol-engined Ford D-series awaits restoration.

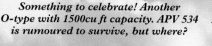

Something to celebrate! Another O-type with 1500cu ft capacity. APV 534 is rumoured to survive, but where?

OLBD in its time with Lidbetter's before passing to Bloomfields in 1955. Note the steeply raked front of the body, which was later rebuilt for more loading space.

Early days. Founder John 'Jack' Bloomfield with brother George and 1929 27hp Chevrolet LQ. The '482' still survives as the second half of the firm's phone number today.

on my feet as far as getting parts was concerned."

Steve's aim had been to get the Bedford finished for this year's Bedford Gathering and he did just that, ADX performing admirably on the journey and nudging 16mpg!

The Bedford has also been reunited with its main driver for many years, Reg Baldwin, who worked for Bloomfields for many years and kindly attended our photoshoot.

"It never broke down once. It would sit there all day at 30mph," he remembers. "I couldn't believe it when Steve said he'd restored the Bedford. I thought it had gone forever. It's just how I remember it, except it now has flashing indicators. I could have done with those."

Reg recalls trundling down to Minehead in Somerset with colleague Jack Orvis. "We had

to put the furniture in a three-storey house, and it took all day. When we'd finished, we each got a bottle of beer and neither of us could drink it. We were so tired."

Stanley Bloomfield is also pleased to see the result of his grandson's determined efforts. "It's good to see it back on the road, and though I didn't physically contribute to the restoration, I gave advice wherever I could. I'd maintained it all those years. My only complaint is that there was a lot of difference in price between what I sold the Bedford for and what Steve paid to get it back. Not that I didn't approve!"

Steve is now embarking on the restoration of the family's 1966 Ford D-series, but it's the Bedford which is the real family heirloom.

The Bloomfields are Bedford fans – indeed the business was built on them. "Apparently at one point before we got Bedfords, the furniture shop used a motorbike and sidecart for deliveries. Not a sidecar, a sidecart. This was used to deliver wardrobes."

Stanley recalls one Bedford-related disaster, which can't

be blamed on the manufacturer.

"We used to move policemen quite a lot and at the time, policemen always had a small farm at the back of the house. We had to carry all their equipment, such as wheelbarrows and even sheds.

"I was going into Newmarket and the bottom fell out of the van. All the furniture literally fell on the road. There was nothing wrong with the floor. It was merely that we were using a lightweight vehicle. There was nothing we could do but stop and wait for my brother to come out to me with another van."

Did the policeman ever know? "Yes, he knew alright, and he claimed compensation up to the eyebrows."

The last word goes to Steve: "This Bedford is not only such a major part of our family's history, but also that of the local area. So many people remember it, and I sometimes can't believe it's finally back on the road."

Bloomfields can be contacted on 01394 282482
Website www.jbloomfield.com

Bedford blends with its surroundings!

Two from the past. PV 3699 was a Bedford W fitted with postwar O-type front, and ERT 260, an O-type bought new.

The fruit-growing area around Hastings is home to a quite amazing selection of TK, A, S and even O-types that see seasonal work carrying fruit from orchards to packing houses and cold stores. A fine example is this splendid rust-free Model A five-tonner, seen loaded with apple boxes at Taradale.

Until the introduction of the famous CA van in 1952, the smallest model in Bedford's post-war range was the PC, a design based on the pre-war Vauxhall-12 care. This excellent and very original 1950 PC utility was seen working for a smallholding near Napier.

KIWI CLASSICS

The famous wartime four-wheel-drive Bedford QL saw military service in all parts of the world, New Zealand being no exception. This very complete example, altered little from the day it left the factory, retains its New Zealand army serial number, despite standing for some years on a dealer's site near Hamilton.

(Photos: Martin Perry)

Our new, regular look at lorries abroad begins with a series by Martin Perry on British classics still working. This month, he's come up with a fascinating collection of Bedfords.

Time was when the famous Bedford advertising slogan, 'You see them everywhere', held true – but Bedfords on the roads of Britain are now a rare and dwindling sight.

On the far side of the world, however, it's a very different story for the Bedford marque is still a dominant part of the everyday road traffic scene in New Zealand.

The 1950s and 1960s were boom times for Bedford – and it's certain that New Zealand was probably the most important export market for the full range of vans, lorries and bus chassis.

Even today, New Zealand's school bus system remains largely in the hands of the Bedford J-type, VAS and especially SB, with many petrol-engined examples still hard at work, often converted to LPG or re-engined

with Isuzu diesel power!

The following selection of vehicles was taken in March this year, and show just a small sample of the Bedfords I came across during normal travel on Kiwi roads – proof that down under you really can still 'see them everywhere.'

Caught by chance one Sunday morning on a quiet road south of Taumarunui, EH 5979 is still operated on a full range of light duties by a local farmer. It is a 1953 model KD.

Until the era of Japanese imports, the little Bedford J1 was a vital part of the Kiwi road scene – but there are plenty still active, with dozens more abandoned in fields and gardens throughout the country. A pristine 1969 example, its 214 petrol engine now converted to LPG, DI 6230 heads for work through early morning traffic in Hamilton.

TKs are plentiful throughout New Zealand, fitted with either petrol or diesel engines. In typical New Zealand sunshine, GU 1303 trundles through Waverley loaded with straw bales.

Bedford's final fling was the TM range – and for the Kiwi market, the company certainly went out in style! The eight-wheeler TM was a New Zealand speciality, but is now becoming hard to find on line-haulage work. This imposing machine, powered by a 6V Detroit two-stroke engine, is very active in the service of Peter Thompson Concrete at Whangamata.

The O-type was once the backbone of New Zealand road haulage. Another former member of the armed forces, dating from 1952, ED 2446 saw service with the New Zealand Air Force until the 1970s. Demobbed with less than 15,000 miles on the clock, the lorry is in immaculate condition – without having been restored. Now privately owned, it's seen in Palmerston North.

A section of readers' recollections on Bedfords, inspired by recent articles in Classic and Vintage Commercials.

IT STILL EXISTS!

The former Bloomfields Bedford feature on the Felixstowe based removals firm is alive, well and restored. APV 534 was a sister vehicle to the beautifully restored OLBC pantechnicon, ADX 486, which we featured on the cover of the November issue.

Mr CF Matthews, Hertfordshire, updates us. He writes: APV 534 is still about and running well, as seen in the enclosed photo.

This vehicle was built on an OWB 8830 chassis, with engine number OB 48594 and OB four point mounting cab, as detailed on the vehicle's buff logbook.

The chassis is stretched 36 inches, with a further 48 inches well at the rear, making it quite long.

The all alloy body was built by Arlingtons onto the existing chassis in 1951, the second example to this design. The Bedford was built for British Xylonite (I think that's the spelling) of Manningtree, Essex, and registered over the border in Suffolk on August 21 1951. Underneath, everything is in good condition, the body overhang keeping most of it dry. The doors are 3x2 oak covered in alloy, with Bedford lift windows - again very solid.

The lorry has still got its vacuum windscreen wiper which works very well - downhill!

The Bedford passed to Bloomfields on August 22 1956. Perry's purchased it from a Mr Green and did most of the work over several years. I finished the work off before we sold it to Dave Stone ('Mr Bedford') at Aylesbury to join his Pickfords pantechnichon. We lost half the chrome from the front while on tour from Haverhill to East Herts.

Like Bloomfields, my father also had a six-wheel Chevrolet in 1930-1, and on one occasion he was travelling along when his rear wheel overtook him!

The article brought back memories of driving similar vehicles for Harry Lawcome, from Whitby, who tells us his story.

What a great story in November's issue about the Bloomfields of Felixstowe Bedfords. It took me back to the 1960s when I drove similar Bedfords for Edwardes Transport, based in Ealing, West London.

Edwardes (I think it was spelt this way) was an old family removals firm that was much smaller in its early days around the First World War, and the office was alongside Ealing Broadway Station. The garage was down the road in a back street near the fire station, and the firm had a furniture store in Acton. We had our own hand petrol pump, and could squeeze three lorries in the garage at a push!

The Bloomfields Bedfords seem to be an almost identical story as we had an ex-army (I think) wartime Bedford rebodied as a furniture van, and a late 40s OLBC which did sterling service without breaking down.

We lusted after new lorries such as the TK, with modern heaters and two wipers, but it wasn't to be! A third lorry was bought, but it was an old Austin K4 with a massive pantechnicon body and four-cylinder engine, which was totally overwhelmed. In fact it wouldn't go up some steep hills when fully loaded!

I don't suppose the firm still exists, and I have no photos of the vehicles. So if any readers have photos and/or information on Edwardes Transport I would love to see them. Allied Ironfounders and Joe Lyons heavy transport at Greenford, Middlesex are two other firms I drove for in the 1960s and again, if anyone has photos of these firms and their lorries in that era, it would make my day!

(Can anyone help - there's pages waiting for a feature on any of the operators Harry mentions - Ed).

Above: The Bedford after restoration, and Left: as it was after years of storage at Bloomfields' premises.

AND THEN THERE WAS ONE.... AND MORE.

More on Bedfords and the much debated in recent magazines issue of originality, written from personal experience by Andy Ballisat from Lincolnshire...

During 1987 I bought a Bedford OLAC of 1950 from Reg Edwards, who had purchased it at auction during 1964.

This had originally been an MoD vehicle and only became registered on Reg's acquisition, so from that point of view had lost its originality.

I added to this demise of originality by the replacement of the body with an ex-MoD Bedford TJ's Luton body.

I've always been inspired by pantechnicons in the livery of Reeds Removals of Sawbridgeworth, Herts, and this is the only way I could fulfil my dream of owning such a vehicle, and the model I've chosen, the OB, is my favourite.

I do feel though that a vehicle in its own original configuration and livery is good to see, but that's not always possible. I didn't want to restore the Bedford in the MoD green.

I'd also like to comment on a similar issue, which could be the subject of an article entitled 'And then there was one.'

McMullens Brewery of Hertford bought 13 brand new Bedford KM drays during 1980, their registrations being OLR 404-416W.

I started taking an interest in their appearances on delivery duties in and around my home town of Sawbridgeworth during the early 1990s. Already they were over ten years of age.

I contacted McMullens during the mid-90s to get permission to visit their garage and workshops in Hertford.

I was pleased with the response, and two weeks later found myself in the care of the workshop manager Richard Hoare, who guided me around the premises, where three KMs were stabled that day.

Apparently, by then only five had survived, with the other five having been cannibalised to donate parts to the remaining fleet still in use.

In the late 1990s I learnt that McMullen's only had three KMs in use, and one was soon to be replaced by another Ford Iveco curtainside dray.

I contacted their office to express my interest in saving one from the same fate as the others, and in May 2001 became the owner of OLR 416W. Coincidentally, this was the same KM I was shown over on my visit to the depot in the mid-1990s.

On collection, I was advised that all signwriting would be erased, and reinstatement wouldn't be allowed.

I have enclosed a photo of this KM when in use as transport for my garage and shed contents to our new home recently.

Sadly, OLR 412W was stripped for spares only a few weeks later, although I am glad to say that 415 will stay for a long time yet as it has been subject to a full restoration.

McMullens will retain this for promotional purposes only, hence the titled 'Then there was one.'

O-Type photographed by Andy on the day he bought it.

OLR 412W was stripped for spares shortly after it was photographed here, but 415 has now been restored for promotional purposes by McMullen's Brewery.

Now with pantechnicon body, outside Andy's then home in Hertfordshire during the late 1980s.

The Bedford KM now owned by our correspondent Andy.

RESTORATION CONTRAS

REMOVALS & WAREHOUSING

P. Taylor & Son
REDHILL SURREY

P. Taylor & Son

REDHILL SURREY

AYT 640

SUP 71E

Main photos: Glyn Burney
Restoration photos: Paul Taylor

One's a huge beast, one's a small slow plodder with wonderfully musical gear whine. Paul Taylor has restored two classic lorries from opposite ends of the scale, a 1967 Foden S24, and a 1934 Bedford WLG. Nick Larkin reports.

Imagine, and this may not be easy, that our collective stock of preserved lorries was a box of chocolates. And you could have two of them. Much as hazelnut whirls might be your favourites, many would feel they were missing out by having a pair of those. They'd possible go for one whirl and something totally different, like a gooey strawberry créme.

Paul Taylor, who runs his own removal firm in Redhill, Surrey, has definitely gone for the contrasting approach with his pair of preserved lorries, which could not be more different in age, style or driving experience.

To hark back briefly to the confectionery theme, one of Paul's vehicles is nigh on impossible to cite in chocolate terms. If you must, its equivalent would have to be a rock 'ard piece of toffee, covered in bitter plain chocolate. Or more appropriately, a gobstopper, or Uncle Joe's Mint Ball. In other words, Paul's 1967 Foden S24 eight-wheeler wouldn't be something you could crunch on easily.

The second vehicle is far more of a fondant fancy, easy on the jaws but still with a enjoyable taste of its own. It's a 1934 Bedford WLG removal van.

Seeing the two now immaculate vehicles together can do nothing else but make you smile. The huge, wide, green Foden towers above the relatively tiny and narrow Bedford.

They could almost be Tom and Jerry.

Even getting in the cab of the Foden is a mountaineering feat, but with the Bedford you just clamber in without thinking, just as you would a Transit van.

Start up both vehicles, and the Foden dominates, its Gardner 180 engine throbbing away purposefully and the whole vehicle shaking in a 'Let's get on with it' way. The Bedford glugs quietly to itself on tickover.

The Foden moves off with an almightily thunderous roar, whereas the Bedford, in its own way almost excels with a range of musical sounds that only 1930s petrol-engined lorries seem to make - a noise - which seemed to find itself on the soundtrack of British films for 30 years when a lorry noise was needed.

Enough of these comparisons - let's find out about the restorations.

98

Rear view emphasises Foden's length yet further!

*Little and Large - 1967 Foden S24
eight-wheeler dwarfs 1934 Bedford WLG*

although instruments relatively comprehensive.

Even the Foden's fuel tank

Bedford cab much like of that of contemporary car.

No unnecessary concessions to luxury in Foden cab...

The Foden

New to FH Harrison of County Durham in 1967, Foden S24 eight-wheeler SUP 71E covered more than a million miles before entering preservation. "I've always liked heavy trucks and this seemed the perfect vehicle," Paul says.

The Foden is believed to have spent some time as a powder tank, before being converted into a flatbed, this work including an extension being welded to the chassis enabling the new body to be some 18 inches longer than the old.

The S24 had been owned by Mike Bonner of Aberdeen for 16 years before he decided to advertise it for sale in 1999.

HGV driver Kevin Jones flew to Scotland to see the vehicle, and ended up driving 600 miles back to Surrey. He'd later carry out much of the restoration.

"Kevin made the journey without problems, but the drive was hard work," said Paul.

Once back home, a leaking rear hub seal was repaired, and a couple of oil leaks rectified.

The engine was also thoroughly serviced, this work being carried out by Albert and Wayne Gilham of W Gilham Services in Oxted, Surrey. This firm also carries out all of the work on Paul's modern fleet.

The Foden's body was cut back to original length: "This was partly so we could get it in our workshops, Paul explained.

Bodywork was now the priority. The flatbed was in a fairly bad state, and parts of the metal framework supporting it were rotten. The cab needed some major work on the glass fibre.

Both the cab and body were removed. Thankfully the chassis was in excellent condition, only needing sandblasting and painting - two-pack being chosen for this.

The cab roof was particularly bad, especially along the moulded strengthening sections, and this work took some considerable time, and an even more considerable amount of glass fibre filler!

Attention was then turned to the flatbed

supports - the side raves (perimeter) were good enough to be salvaged, but the cross sections and runners were made in hardwood, supplied by a timber yard next door to Paul's Redhill premises. For the actual flat, hardwood garden style decking was used - and the results look superb. Even Alan Titchmarsh would be impressed!

After various cracks were repaired, the cab was repainted, again in two-pack.

The Foden's electrics were checked and found to be in good condition. The brakes, which had been stripped down when the chassis was sandblasted, were cleaned and checked

Paul had been determined to get the Foden ready for the 2000 London-Brighton Run. "A week before, the body still hadn't been attached or so we were working past 10pm every night. At least the dog did well out of this. It got my dinner every night!"

gnwriter's art!

Mighty Gardner 180 engine in Foden.

Foden maker's plate remains intact.

Paul Taylor (centre) with driver/manager Spirit says both Foden and Bedford have strong individual appeal.

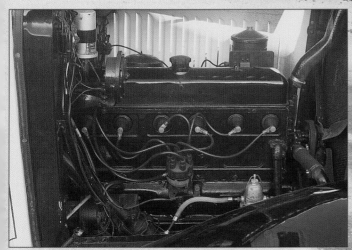

Bedford's six-cylinder petrol engine gives reasonable performance.

Foden and Bedford both named after members of Paul Taylor's family.

We've spent ages wondering - do these headlights come from a car - and is it the Vauxhall Viva HB?

Doors removed and cab body ready for full inspection.

Lift cab! Strengthening ridges on roof needed major repair work.

Windows, lights and trim were installed just in time and at 4pm on the day before the run, signwriter John Edge from East Grinstead, whose work also appears on the Taylor's modern fleet, finally finished weaving his magic on the Foden.

"Then we were working up to 10pm that night, and up at 6am for the finishing touches before driving to London for the start of the run, getting used to the Foden's 12-speed gearbox en route. "Well, that's if you ever do get used to that gearbox," said Paul. Must be some selection box chocolate analogy there!

Anyway, back to Paul: "Still it was a great day, we were made very welcome and the Foden looked like a new vehicle."

The Bedford

New in 1934, the Bedford WLG is believed to have spent time working in locations as diverse as Ireland and Derby.

Paul bought it from the John Mould sale - the still much talked about event in which John was forced to sell off some of his huge collection of classic lorries thanks to the local council.

"I had always liked Bedfords - we used them from the beginning in the business, and it seemed a good idea to have a removal vehicle for promotional purposes," Paul explained.

Paul Taylor Removals is now celebrating its 21st year. Paul had been tiring of his job doing shift work with East Surrey Water as a store keeper/lime runner. He bought a secondhand Bedford CF van, NLP 575L, in 1980.

TIPS FOR RESTORERS

1 Don't take on a project unless you know you can tackle it.

2 Don't forget the size of the vehicle - can you accommodate it.

3 Plan the restoration carefully, and don't forget to be ready for the unexpected!

Foden's cab needed major glass fibre repairs.

Apt headline in local newspaper report of vandalism incident.

"I began collecting waste cardboard and selling it for £6 a ton," recalls Paul.

This expanded into removals, and Paul decided it was time to give up the day job. The result is a company with 22 vehicles - all named after family members of Paul and his staff, 10 full-time employees and half a dozen part-timers.

With two depots at Merstham and Redhill, Surrey, the firm not only specialises in removals but also skip and grab hire. Self drive lorries and vans, along with removal kits are available to hire.

Bedford CF Luton vans and TKs played a major part in expanding the business. "Bedfords have always been very reliable and I have a lot to be grateful to the marque for," said Paul.

"The WLG ran well, but some bodywork was needed and it was painted in an awful brown colour," he recalls

Once back in Paul's workshops, the vehicle was fully inspected. Attention was needed to the A-posts, which needed careful welding.

This completed, the Bedford was treated to a repaint, again in two-pack, and, like the Foden, was professionally signwritten.

A good service and tune-up for the engine, and brake adjustment were virtually all that was needed before the Bedford was ready for the road

The WLG completed the 1999 and 2000 runs before, on the 2001 event disaster struck.

Disasters... and solutions

Two of the major setbacks most feared by classic commercial owners have landed

Bedford gets the paintshop treatment.

slap bang on Paul's lap.

Firstly, both the Foden and Bedford were attacked by verminous vandals in a car park near Taylor's now vacated premises in Oxted, Surrey.

Major damage was caused to the Foden's paintwork after it was attacked with a Stanley knife or similar. Said vermin also broke into the Bedford, but managed to confine their attentions to an old bench in the back of the vehicle.

"It was disheartening. The Foden looked like a new lorry when this happened. You don't just restore these vehicles for your own pleasure but also that of others. Then this happens and you're tempted to hide the vehicles away," said Paul.

The Foden still bears some signs of its attack, but Paul has done a good job of disguising the damage as soon as possible.

Adds Paul: "I was gutted when this happened, and the attitude of the police didn't help. They didn't even come to see me."

Paul claims that this was despite witnesses who came forward and said they'd seen certain individuals running from the scene.

The Bedford was to have its own catastrophe, when the engine blew up on the 2001 London-Brighton Run. "We believe there was something wrong with the oil pump, which resulted in the big ends being knocked out," Paul recalls.

The engine was rebuilt by Albert and Wayne Gilham, a new clutch and other parts being sourced by regular *CVC* correspondent Norman Aish, of Bygone Bedford Bits fame. The WLG now runs superbly.

So Paul's contrasting duo are ready for action at classic commercial and local charitable events.

So how does he sum them up? "The Foden's a beast, a real beast, and I love driving it, though with the suspension you tend to spend more time out of the driving seat in the air than sitting on it. The Bedford's a slow plodder, and very much an 'old' lorry but it's good fun, more manoeuvrable than the Foden... and there's that gearbox whine."

> Many thanks to Paul and his family for all their help with this feature - and also transport manager 'Spirit' for his assistance.

Bedfords have played a major part in Paul's business development.

Different fronts... and talking heads!

The leader! Tony Knowles's 1974 KM won the [...] behind is a six-wheeler VAL coach – a chassi[...] at the gathering.

(Photos: Nick Larkin, except where stated)

BEDFORDS E

More than 200 vehicles assembled for this year's Bedford Gathering – and what a collection.
Nick Larkin reports.

Everywhere you looked, you saw a Bedford, from a TK to a CA to a coach. Venerable, vintage, almost modern; there was something for every marque-related taste at this year's Bedford Gathering. And many real gems. There were even Bedfords from abroad – and a good display of non-Bedfords – during the event at the Holiday Inn, Cambridge.

Voted Best Preserved Bedford was, for the third year running, the Wychwood Brewery from Oxfordshire's 1951 Bedford OLBC, GSJ 761 – though the owners hadn't intended for it to be judged. Not that they weren't pleased to win the treble!

Talking of O-types, an extraordinary six-wheeler made its Bedford Gathering debut in the form of LMV 550, a 1946 OLBD dropside owned by Clive Matthews of Bishop's Stortford. The lorry was new to locally-based Frederick

Catton and Sons, who commissioned several similar conversions.

Clive, a well-known O-type collector, has carried out a major restoration on the six-wheeler. He revealed that Catton had work done at the former Scammell & Nephew works in Fashion Street, London. However, no one is certain whether the conversion on the Bedford was carried out there.

Another attention-grabbing rarity never seen at a vintage vehicle event was John Kilby's 1956 Bedford A4 horsebox with Lambourn coachwork. John, from Hertfordshire, bought the vehicle in 1985 for use as a spare horsebox. It had been used sparingly until 1991. "It's very slow but characterful," was how he summed up the vehicle. He was also exhibiting 1952 Bedford 0-type horsebox with Vincent body, ERD 859.

Moving forward three decades and the Ellesmere Port-based Griffin Trust was displaying its 1985 Bedford TM artic and trailer, B923 UBM, which has been converted into an exhibition unit.

Latest in a superb line of restorations carried out by Tony Knowles of

Wimblington, Cambs, is 1974 Bedford KM artic, WFE 711M, voted Best Bedford of the Show.

Lightweight Bedfords were, of course, a major attraction. Among them was the 1962 Bedford CA with Martin Walter Utilibrake conversion owned by equally light(?) *CVC* publisher Peter Simpson – Kelsey Publishing were among the sponsors of the event.

The Bedford VAL twin-steer chassis also celebrated its 40th anniversary. Although designed for coaches, they were sometimes used for commercials, such as mobile television broadcast units. We counted 18 VALs – surely the biggest preservation gathering ever.

Bill Wood's wonderful 1946 Bedford M-type, which made its debut at our Donington Show in March, has now been joined by a K-type DTW 350, in similar flamboyant but original fruiterer's livery. Watch out for the story on them both, coming soon to *CVC*.

A sizeable contingent of rally entrants from abroad were among the participants, including long-distance award winner Chris Balkenende, from Edam, Holland, with his 1976 Volvo F88.

Talking of vehicles from long distances away – Paul Richman was showing a 1949 Bedford K re-imported from New Zealand. He is hoping to bring many

Watch out for the full story in CVC coming soon. Bill Wood's Bedford M-type has been joined by superb K, DTW 360.

...rd award. Just visible ...s its 40th anniversary

Close your eyes if you're of a nervous disposition. It's CVC's publisher Peter Simpson with his recently acquired 1962 Bedford CA.

Major prizewinner three years in succession, Wychwood Brewery's 1951 Bedford OLBC.

VERYWHERE!

Long Distance Award winner was this 1976 Volvo F88, owned by haulier Chris Balkenende of Edam, Holland. It was new to a graphic machine manufacturer in Gothenburg, Sweden, and has covered under 325,000km. The Volvo was transporting Chris's other show exhibit, a 1957 Bedford RL 4x4, new to the British Army on the Rhine before passing to a Dutch haulier.

Yet another rarity, 1956 Bedford A4 horsebox with Lambourn coachwork.

The Griffin Trust has converted its 1985 Bedford TM artic and trailer, B923 UBM, into an exhibition unit, which can be hired in aid of trust funds. Details on 0151 327 4701.

(Photo: Fred Jerabek)

Above: Bedfords – and some vehicles not even Bedfords.

FRED^K CATTON & SONS

(Photo: Gyles Carpenter)

Still delivering coal to the citizens of West Auckland, 1981 Bedford TK of Douglas Mairs and Sons was voted Best Working Vehicle.

Cheers! See you next year! A friendly wave from Joe Ovel of Somersham, Cambs, as he heads for home in this 1936 Bedford BYC. It took the award for the oldest example of the marque.

other classics back from Kiwiland. See News for further details.

Organisers' spokesman Dick Haughey said: "It's been a fantastic event – we've had a greater variety of vehicles this year than ever before."

Surprise of the show? We haven't doctored the photo – this really is a six-wheeler 1946 OLBD dropside owned by Clive Matthews of Bishop's Stortford, who has carried out a major restoration. The lorry was new to locally-based Frederick Catton and Sons, who commissioned several similar conversions.

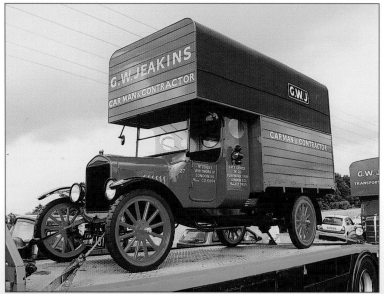

Lightweight winner, wonderfully restored 1921 Model T loaded for home.

WINNERS

Best Preserved Bedford: GSJ 761, 1951 Bedford OLBC, Wychwood Brewery

Best Bedford: WFE 711M, 1974 Bedford KM, artic, Tony Knowles

PSV Still in Service: UDY 910, 1987 Bedford YMPS/Plaxton, C Rowland. Hastings

Light Van/Pickup: IW 181, 1921 Ford Model T, HW Pickerell, Billericay

PSV 20 seats or Under: KNK 373H, 1969 Bedford J2/Plaxton, Kenzies Coaches, Shepreth

Best Working Bedford: RDC 531X, 1981 Bedford TK, Mairs, County Durham

Non-Bedford PSV: AJD 959, 1945 Morris Commercial CV11/40, Stocker body, Felix Taxis, Long Melford

Furthest Travelled: BL-HG-62, 1976 Volvo F88, Chris Balkenende, Edam, Holland

Best Under 3 Tons: DTW 350, 1952 Bedford K Type, B Wood, Waltham Abbey

Best Bedford VAL: SNT 925H, Plaxton C53F body, 1970, Andy James, Tetbury

Oldest Bedford: DCD 567, 1936 Bedford BYC, J Ovel, Somersham

Bedford OB: DBU 889, 1947 Duple C27F-body, Felix Taxis

Bedford PSV: 1951 SBG/Duple C33F, G Heels, Ashtead

Best non Bedford HGV: RTS 457, 1973 ERF 54.G, flatbed, J Richards & Sons Ltd., Fakenham

CF cluster among smaller Bedfords.

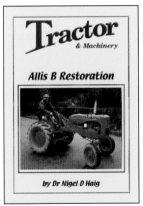
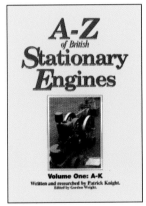
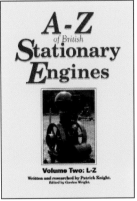

TK TRIUM

This rare Bedford TK tipper has been restored in British Racing Green, and is creating a lot of attention on the rally circuit. Alan Barnes reports.

(Photos: Alan Barnes)

Prior to the Second World War, George Brennan worked as a driver for WG Hinton & Sons Ltd, a Southampton-based building firm. With the outbreak of hostilities, George found himself in the army. He was obviously held in high regard: Mr Hinton wrote to him while he was away serving King and country, urging him to return to the firm when he was de-mobbed. George treasured that letter, which he kept for the rest of his life – yes, he did re-join the firm after the war and stayed for many years.

In 1972, the company ordered a new tipper from local Bedford dealer, Picador Motors of Portsmouth Road, Sholing. George collected the addition – a Bedford TK 220 diesel, DOR 389K. Bedford produced this version for only one year, which is distinctive in having the front indicators located down by the cab steps. George would not allow anyone else to drive it, and so the Bedford became known as 'George's Lorry'. Sadly, George died a few years ago.

As for the Bedford, well, it was stored in a barn for nearly 15 years with just over 66,000 miles on the clock. In 1988, the TK was bought by a R Stillo, who kept it for only a few weeks before putting it up for sale in a local newspaper. John Appleton, a vehicle technician from Slough, paid £850 and began the restoration almost as soon as he got it home.

Now 'George's Lorry' is back on the road and looking magnificent. I met John earlier this year on the HCVS-organised Ridgeway Run in the far-from-picturesque surroundings of the station carpark in Henley-on-Thames. The Bedford looked immaculate in its new British Racing Green livery, but the overcast conditions meant that any photographs would hardly do the vehicle justice. What we needed was some sunshine, so we agreed to meet later in the summer.

Call me a fair weather photographer if you like – and Sentinel owner Malcolm Rogers certainly does – but I feel a

spot of sunshine always brings out the best in any vehicle.

John and I arranged to meet one Saturday in August, and as luck would have it, that day turned out to be one of the hottest of the year with temperatures of 30 deg C recorded in London. John had sourced one or two locations around Slough beforehand, and the site at the Riverside Café on the A4 at Colnbrook proved ideal. Resisting the temptation of bacon and eggs, we somehow managed to concentrate on the job in hand.

John obviously has Bedfords in his blood. His father, also called John, drove them as an owner-driver in the 1970s. There is a delightful picture of a two-year-old John standing in the cab of his father's lorry. "I've

Gleaming in British Racing Green, new to WG Hinton Bedford TK tipper.

DOR 389K

always been a TK fan. They're easy to drive, it's easy to get into the cab and they were well made," he says.

Leaving school at 16 and spending three years as an engineering apprentice, John joined Mercedes-Benz when he was 19 and worked there for four years. After a stint elsewhere, he returned to Mercedes a couple of years ago, working as a technician.

John's engineering training has been fully used in restoring 'George's Lorry', a project completed almost single-handedly to an exceedingly high standard.

First on the agenda was to get some welding done, most importantly fitting a replacement roof. "It had rotted around the drainage channel. I managed to get hold of another TK which was unrestorable – the chassis was in a terrible state, but the roof was in good condition, and this was welded in," said John.

Thankfully, the chassis was good and only needed shotblasting before repainting. But this did not apply to the body. Recalls John: "Being a builder's truck, the cab and chassis were well maintained, but the body had so many dents it was irreparable." The replacement body had been on a Mercedes 814, but was virtually identical to that seen in a Bedford TK

brochure. Thankfully, the tipper ram and PTO needed no attention.

The wheels were shotblasted. A new exhaust system was fitted and rear wheel cylinders, brake shoes and hub seals were replaced. The original Bedford 220 diesel engine was in pretty sound condition, but required new gaskets along with a thorough service to bring it up to standard. Front engine mountings were renewed, the alternator re-conditioned, fan belts replaced and a new radiator, hoses and water pump fitted.

A Bedford dealer came up with original wing mirrors, and two rear mudflaps from a 1978 TK were fitted along with a new toolbox. The list seemed never ending – it is hard to appreciate that almost all this work was done at John's home.

Following road testing, the original 5.7 differential was changed to 4.3 ratio, and the Bedford can do 65mph quite happily. I wonder what George would say about that! John explains why this was done.

"When I started going to some rallies further away, it wasn't much fun travelling at 45mph with lorries coming up behind

you and not realising the speed you were doing until the last moment. Being built as a tipper, it was very low geared.

"Fitting the new diff was quite a task. It had to have the bearings shimmed and everything – but it was well worth the work. The TK would do around 17mpg with the original diff – now it'll do 23."

The re-building was to take the best part of three years and following the mechanical restoration, the Bedford went to CCR in Iver, Bucks, for painting. So why British Racing Green when the lorry was originally in blue?

"Well, I've always liked TKs, and I've also always liked British Racing Green. So why not? Besides, I've always imagined TKs with black mudguards, so this one has got those, too."

The TK was then sent to Saunder's Signs in Watford for signwriting. The artwork includes the names of John's wife, Carla, and his two daughters, Lauren and

Bumper boasts the names of John's family.

TK with low-mounted indicators was only produced for a year – and tipper variants are rare.

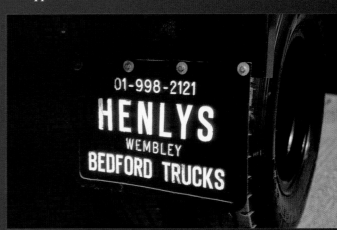

01-998-2121
HENLYS
WEMBLEY
BEDFORD TRUCKS

Mudflaps came from Bedford dealer – note what is now a 'period' London phone number.

Power take-off (PTO) and tipping ram were in good condition, as was the chassis.

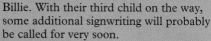

Appropriate signwriting contrasts well with the green for overall effect.

Billie. With their third child on the way, some additional signwriting will probably be called for very soon.

One problem surfaced after the initial repainting of the replacement tipper body. "It didn't look so bad in unpainted aluminum, but once it was British Racing Green, you could see all the dents. In the end, a friend and I made new body sides and these were then resprayed," said John.

Another problem occurred when the alternator was overhauled and a new bearing fitted. "Something happened to the regulator, which meant it was putting out 17 volts. It nearly melted the battery."

Although there are one or two minor jobs on the interior to finish, the Bedford is largely complete. It took part in the HCVS London to Brighton Run last year, the Ridgeway Run last June, and several rallies up and down the country.

After tracing the TK's history via the

DVLA, John managed to track down Mr Hinton, now in his mid-eighties. "He was over the moon that the Bedford had been saved and restored," said John. "When I told him it was in British Racing Green, he pointed out that the firm's headed notepaper was always green and cream – so that was a coincidence."

So what does John feel like tackling next? He's admitted to looking at one or two other vehicles, but as they needed very little work, they had not captured his interest the way 'George's Lorry' had. He says: "If I had the money, I'd definitely do another restoration." Sounds like we'll be hearing from John again!

THANKS

Thanks to the Riverside Café for allowing us to use their lorry park for the shoot. Thanks also to John for not only for putting up with my endless requests to

move a few feet here and a few feet there, but for taking the time to search out some of the history of George and his Bedford.

JOHN'S TIPS FOR RESTORERS

1 Don't do it every weekend – you'll get bored. Better to work every other weekend instead.

2 Don't get disheartened. You'll get it done in the end!

3 Shop around for spare parts. You can get some from main dealers, but they are often cheaper from other specialists or at rallies.

Bedford on the road – and capable of 65mph and 23 mpg!